Njikam Aboubacar Sidik Lacatus
Seino Richard Akwanjoh
Taku Awa II

Changes in vegetation cover and habitat deterioration

Njikam Aboubacar Sidik Lacatus
Seino Richard Akwanjoh
Taku Awa II

Changes in vegetation cover and habitat deterioration

White-throated Babbler (Kupeornis gilberti) in Bakossi National Park

ScienciaScripts

Imprint

Any brand names and product names mentioned in this book are subject to trademark, brand or patent protection and are trademarks or registered trademarks of their respective holders. The use of brand names, product names, common names, trade names, product descriptions etc. even without a particular marking in this work is in no way to be construed to mean that such names may be regarded as unrestricted in respect of trademark and brand protection legislation and could thus be used by anyone.

Cover image: www.ingimage.com

This book is a translation from the original published under ISBN 978-620-6-71110-0.

Publisher:
Sciencia Scripts
is a trademark of
Dodo Books Indian Ocean Ltd. and OmniScriptum S.R.L publishing group

120 High Road, East Finchley, London, N2 9ED, United Kingdom
Str. Armeneasca 28/1, office 1, Chisinau MD-2012, Republic of Moldova, Europe
Printed at: see last page
ISBN: 978-620-7-74161-8

DEDICATION

I dedicate this work:

- To my father Moungnutou Ibrahim and;
- To my mother Peyouonmiem Ramatou, may she rest in peace.

ACKNOWLEDGEMENTS

The preparation of this document was made possible thanks to the help and support of many people. I would like to thank all those who have contributed their physical, intellectual and emotional efforts to its realization.

First of all, I thank Almighty Allah(God) for the breath of life, His love and faithfulness.

My immense gratitude to all the academic and administrative authorities and all the teaching staff at the University of Dschang for the quality of the guidance I have received over the past five years.

My sincere thanks:

- To my supervisor Prof. Seino Richard Akwanjoh, who agreed to lead and direct this work from start to finish. His human qualities and love of a job well done command respect and admiration. May he rest assured of my gratitude;
- Dr Taku Awa II, lecturer in the Department of Animal Biology at the Faculty of Science, University of Dschang, for his personal commitment and support;
- All the teachers in the Department of Animal Biology, to whom I am grateful for their various contributions to my training;
- Ngouh Amadou, PhD student at the Faculty of Agronomy and Agricultural Sciences, who helped me obtain the Landsat satellite images;
- To my academic elder Francis Guetse, who guided me from the data collection phase to the writing of this document;
- To Guilain Tsetago, doctoral student in the Department of Animal Biology, who made a major contribution to improving the quality of this document;
- To all my graduates: Wongibé Poupezo Dieudonné, Kenfack Assuna Melerine, Tenonfo Ngouefack Sophie Rella, Meka Mbiézié Mesmine, Petnga Alex Stephane, Tigumung Cyril, Ntene Sob Branly and Fofie Marissa;
- To my field guide Mr Claude;
- To my guardian Mr Nkoma Adamou and his wives: Dawouo Abiba and Lounga Mariatou, peace be upon them.
- To all the children of the Nkoma Adamou family
- To my grandmother Ayiagnigni Maimouna;
- To my mother-in-law Kpouyié Alima;
- To my fiancée Betche Michele Doriane, the mother of my daughter;

- To my daughter Njikam Weladji Nahila Lyana
- To all my brothers and sisters: Liefouyoum zoukifilou, Awouoyiégnigni Chadoure Adamou, Nzinah Adja Alima Ladouce, and Lijouom Afsa Dairou;
- To the people of Bakossi for their hospitality during the field phase;
- To my aunt Yenou Awawou and her children;
- All those who helped me in any way in the realization of this work.

TABLE OF CONTENTS

LIST OF ABBREVIATIONS

BEPC: Brevet d'étude du premier cycle (Certificate of Secondary Education)

BUCREP: Central Bureau of Population Census and Studies

BNP: Bakossi National Park

CARPE: Central African Regional Programme of the Environment.

CBD: Convention on Biological Diversity

CEP: Primary School Certificate

CDC: Cameroon development coorporation

COMIFAC: Central African Forests Commission

EBA: Endemic bird areas

EU: European Union

FAO: Food and agriculture organisation or organisation des nations unies pour l'alimentation et l'agriculture

GFRA: Global forest resource assessment

GPS: Global Positioning System.

JRC: Joint research center EC-JRC (Europeen commision- joint research center) or institut pour la référence matériels et mesures.

MINFOF: Ministry of Forests and Fauna

MINEF: Ministry of the Environment and Wildlife

NGO: Non-governmental organization

PARCC: Climate Resilient Protected Areas

GNP: Bakossi National Park

SCBD: Secretariat of the Convention on Biological Diversity.

GIS: Geographic Information System.

TAGB: White-throated Babbler

IUCN: International Union for Conservation of Nature.

WTMB: White throated-mountain Babbler

SUMMARY

The White-throated Timalie (*Kupeornis gilberti*) is classified as one of the world's least studied "endangered" avian species (IUCN 2014, Birdlife international 2016). It is endemic to the Bakossi National Park (BNP), where it faces threats such as agriculture and habitat degradation. The lack of information on this species drew our attention to the rate of change of its habitat as well as the factors influencing its probability of occurrence due to these threats. During the period from December 2017 to January 2018, data were collected at six sites. At each point, we looked for the presence or absence of the species. In addition, Geographic Information Systems (GIS) and remote sensing enabled us to use Landsat satellite images from 2006 and 2018 to assess changes in vegetation cover from 2006 to 2018, and to classify cover types in relation to changes between 2006 and 2018. The study showed that the remaining population of the White-throated Babbler (TAGB) is distributed mainly in dense forests and often observed in secondary forests. Its probability of encounter increases significantly in primary forest (75%) and secondary forest (25%). However, its probability of encounter increases significantly in primary and secondary forests, but is nil in savannahs and crop fields. During the same period, we noted a 13% loss of dense forest cover (from 79.41 to 66.41%) due to the effect of deforestation, an increase in secondary forest cover (from 17.89 to 31.17%) and human activities (from 0.42 to 0.82%). This increase in surface area is due to the conversion of forest cover to industrial plantations and infrastructure. Finally, we noted a loss of savannah area (from 1.12 to 0.77%) and bare soil (from 1.16 to 0.83%) due to the effect of bushfire. This loss of forest area at low altitudes is justified by the creation of plantations and infrastructure in the GNP belt. The growth of plantations and deforestation are causing the conversion of forest cover to agriculture and the loss of wildlife habitat. To mitigate the threats that are responsible for the loss of White-throated Babbler habitat, we propose: the delimitation of the GNP, the intensification of patrols and the encouragement of perennial agriculture.

Keywords: White-throated Babbler, GIS, remote sensing, canopy, landsat , Bakossi National Park(BNP), Cameroon.

CHAPTER I: INTRODUCTION

1- Context

Among the threats contributing to biodiversity erosion today, habitat degradation and deforestation alone contribute up to 85% to biodiversity loss (IUCN 2016). Among these, 30% of amphibians, 23% of mammals and 12% of birds are threatened (Vié *et al.;* 2008). Protected areas have long been used as one of the key strategies for conserving species and ecosystems. However, they are under increasing threat from climate change, which is now being exacerbated by other anthropogenic pressures (Impact of Climate Change on Biodiversity and Protected Areas in West Africa, PARCC). Direct habitat destruction and degradation through land use and land cover change are among the most significant and immediate threats to biodiversity (Titeu *et al.;* 2016). Agricultural pressures and habitat fragmentation have been the cause of the loss of several passerine species throughout the half of the 20^e century, occurring mainly on the peripheral reedbed whose characteristics (surfaces and isolation of fragments) according to Mariano P 2008.

The forests of the Congo Basin are the second largest tropical forest ecosystem after the Amazon rainforest. Its surface area is estimated at around 200 million hectares, or nearly 91% of Africa's dense rainforests. These forests contain an extraordinary biodiversity that represents an invaluable potential for the socio-economic development of the region (CARPE, 2006; COMIFAC, 2009). At the Yaoundé summit in 1999, the Heads of State of Central Africa agreed to examine the problems associated with the conservation and sustainable management of Central African forest ecosystems (NGONO, 2014). Sustainable forest management and/or protected areas(PAs) is a means of conserving benefits for present and future generations in many countries around the world, but this is not the case due to changing vegetation cover (Nagendra 2008; FAO, 2015). In some countries, the problem is due to the absence of an appropriate forest policy, non-enforcement of forest law and lack of structure: the case of Cameroon. Currently, conservation faces many challenges such as deforestation, soil degradation and poverty (Mukete *et al.;* 2018b). Tropical rainforests are among the most degraded ecosystems on the planet, due to the conversion of forest areas into farmland (Achard et al.; 2002; Stork *et al.;* 2009), particularly in Africa (Lepers *et al.;* 2005; Mayaux *et al.;* 2005).

In Africa, population growth is a major cause of these changes (Bawa and Dayanandan 1997; Bamba *et al.;* 2010), not to mention the growing demand for wood for export, according to Oyono *et al.* (2005). Nowadays, the methods most in demand for ecological studies focusing on species habitat change include remote sensing and GIS (geographic information systems), which enable time-saving, less costly and faster operations, as well as the assessment of landscape change over large areas (Gottschalk *et al.;* 2004; Loveland and Dwyer, 2012; Hansen *et al.;* 2013).

Endemic and threatened species such as the White-throated Babbler (*Kupeornis gilberti*) are highly vulnerable to habitat loss (Borrow and Demey **2004**; Birdlife international 2003; Norrow; hartz). This is the case for this species, which specializes in dense mountain forests (Collar and Stuart, 1985).

Knowledge of the factors responsible for the population decline is of paramount importance for the conservation of the species.

2. Issues

The White-throated Timalie is a forest specialist (Danjuma *et al.;* 2014) classified as endangered according to IUCN Red List criteria (IUCN, 2016). It is endemic to the West (*Endemic Bird Area*) of Cameroon and a small fringe of the population has been observed on the Obudu Plateau (*Important Bird Area*) in Nigeria according to Collar and Stuart; 1985. The literature reveals that Mont Bakossi and the Rumpi hills are sites with a high population density of this species (Dowsett Lemaire and Dowsett 1998d). In 1998, the Bakossi population was estimated at several thousand individuals, with a preliminary figure of 10,000-19,999. But only 6,667-13,333 individuals were mature in the Bakossi forests, according to Collar and Stuart 1985.

Recent studies point to extensive deforestation in recent years on both the Obudu plateau (Danjuma *et al.;* 2014) and the Rumpi hills (Tamungang *et al.;* 2014; Mukete *et al.;* 2018). Faced with the exponential growth of industrial oil palm plantations, coffee and cocoa fields and the exploitation of wood products, bush fires, intensive livestock farming and agriculture according to (Linder *et al.;* 2011), there is cause for concern for a forest specialist species such as the White-throated Babbler.

Although Bakossi is an important site where a large population has been found, no studies have yet been carried out to assess the species' habitat change, distribution, population status and ecology.

This study will assess the level of habitat change and the factors responsible for the deterioration of white-throated blackbird habitat in Bakossi National Park.

3-Research questions

Because of the lack of data on the conservation of the White-throated Babbler, there are important issues of concern in relation to the growth of agricultural plantations and the destruction of forests:

i)   How has the habitat of the White-throated Babbler changed from 2006 to 2018?

ii)   What agents are responsible for the change in the habitat of the White-throated Babbler?

iii)   What habitat does the White-throated Babbler use?

4-Hypothesis

i)   White-throated Babbler ranges are thought to have declined as a result of human activities. Many wild species are affected by human-induced changes in land use that occur on large spatial scales (Mather and Needle, 2000; Geist and Lambin, 2002; Jansen *et al.;* 2008).

ii)   The creation of Bakossi National Park would have slowed the rate of deforestation in Cameroon's Bakossi Mountains, given the role that protected areas play in biodiversity conservation (IUCN/PACO, 2009).

5 Research objectives

5.1 General objectives

This study on the conservation of the White-throated Babbler in Bakossi National Park will assess habitat deterioration and conservation threats to the White-throated Babbler in Bakossi National Park.

Specific objectives

i) Evaluate the evolutionary trend of vegetation cover in Bakossi National Park between 2006 and 2018.

ii) Identify the different types of habitat used by the White-throated Babbler and draw up a vegetation map of Bakossi National Park.

iii) Identify anthropogenic activities in Bakossi National Park responsible for the degradation of White-throated Babbler habitat.

6. Justification for the study

Africa's bird species are globally threatened, but the factors behind the decline in these populations are not known (Birdlife international 2013). The major metropolises of abundance for the White-throated Sandgrouse are the Obudu plateau (Nigeria) and Rumpi hills, Mont Kupé and Bakossi in Cameroon. Endemic forest birds are highly vulnerable to habitat loss (Brooks *et al.;* 2001; Borrow and Demey 2004). The White-throated Babbler is a threatened endemic mountain forest specialist and is classified as such. No information is available on its conservation. The ever-increasing activity in the south-west in search of new land is of particular concern to Bakossi, as population growth forecasts are truly galloping (BUCREP, 2010b), yet this population is essentially dependent on agriculture.

Among a multitude of methods based on satellite images we have: remote sensing and geographic information systems (GIS). Advances in GIS and remote sensing have helped ecologists to assess the influence of spatial habitat configuration on ecological processes (Scott *et al.;* 1993; Johnston 1998, Millington *et al.;* 2002). Both techniques can be used to assess changes in biological diversity (Roy and Tomar 2008; Nagendra 2001), land cover change (Franklin *et al.;* 2000) and climate change leading to habitat loss (Scott *et al.;* 1993; Kerr and Ostrovsky 2003).

Many forest specialist bird species are negatively affected by forest disturbance, and insectivorous birds have disappeared in some heavily transformed forests (Sekercioglu *et al.;* 2002; Chace and Walsh 2006; Gove *et al.;* 2008, Danjuma *et al.;* 2014). Similar studies carried out on the Obudu Plateau in Nigeria and Mount Kupé and Rumpi Hill in Cameroon; two areas of abundance for the species (Birdlife international 2000), revealed that the White-throated Babbler appears to be an insectivorous, forest specialist found on primary forest canopies generally and occasionally in secondary forests (Collar and Stuart 1985). Our aim is to find an area where this habitat can be found. However, the two most important sites for the White-throated Babbler are the Bakossi and Rumpi hills, as these are the areas with the densest forests (Dowsett Lemaire and Dowsett 1998d). This justifies our choice of study site in the Bakossi National Park.

The White-throated Babbler is a globally threatened, sedentary and endemic species (Borrow and Demey, 2004) in Cameroon, and is experiencing a severe population decline. This study on the deterioration of the habitat of the White-throated Babbler will provide us with information on the effects of anthropic activities, the causes of its decline and the different types of habitat frequented by this species.

7. Importance of the study

Our study is of two kinds: theoretical and practical.

❖ Theoretical interest

This study provides us with useful information for the conservation of the remaining population of the White-throated Babbler in Cameroon. It will enable us to understand the different types of habitat frequented by the species, the factors responsible for the deterioration of its habitat and the causes of landscape change in the study area. The results of this study will enable us to adopt initiatives aimed at better conserving the species.

❖ Practical interest

This work provides information that will call on the State and the Ministry of Forests and Wildlife in particular, to set up a management plan for Bakossi National Park, which will be used to draw up patrol programs with the aim of putting an end to the anthropic activities carried out by the riparian population. This work calls on the State to attract NGOs for effective conservation in the GNP.

CHAPTER II: LITERATURE REVIEW AND CONCEPT DEFINITION

I.1 Theoretical and conceptual approach

In this section, we define the concepts and outlines we will be using:

Protected areas according to the updated definition of the International Union for Conservation of Nature: a protected area is "a clearly defined geographical space, recognized, dedicated and managed, by any effective means, legal or otherwise, to ensure the long-term conservation of nature and its associated ecosystem services and cultural values". This concise definition determines the fundamental objectives of protected areas: protection and maintenance of biodiversity (understood in three dimensions: specific, genetic and ecosystemic) of natural resources, countries and related cultural values (Carole M and Patrick T 2012).

National park: area set aside for the conservation and propagation of fauna, wild flora and biological diversity, for the protection of sites, landscapes and geological formations of particular aesthetic value, and for scientific research, education and public recreation (IUCN 1994).

Fragmentation: the dynamic process of reducing the surface area of a habitat and separating it into several fragments by barriers (such as road infrastructure) or by the creation of patches that cannot function as the original habitat for the current species pool. Fragmentation implies both a reduction in the total surface area of the habitat and an increase in the isolation of individual patches from one another. According to Danjuma *et al* (2014); Wilcove *et al;* (1986).

Habitat fragmentation is the process of subdividing a continuous habitat into smaller parts, which occurs in systems. It involves a loss of habitat, a reduction in size and an increase in distance between isolated areas, but also an increase in new habitat (Henrik A, 1994).

Ecosystem according to the Convention on Biological Diversity: "a dynamic complex of plant, animal and micro-organism communities and their non-living environment interacting as a functional unit".

Biological diversity: according to Article 2 of the Convention on Biological Diversity, biological diversity is defined as the variability among living organisms from all sources, including terrestrial, marine and other aquatic ecosystems and the ecological complexes of which they are part; it includes diversity within species, between species and of ecosystems (Pullin, 2002).

Land use: all human activities directly related to the land and affecting its resources. Conservation: the aspect of management which ensures that the use of resources is sustainable and that the ecological processes and genetic diversity essential to the sustainability of the resources in question are preserved (agriculture, fisheries, forestry and wildlife), according to Arnold and Jongma (1977).

Landscape: set of ecosystems that coexist in a geographical area, according to the lexicon of protected areas in French-speaking Africa. Deterioration: the action of damaging, spoiling or making worse, according to the Larousse dictionary.

Land cover: physical, biological and chemical categorization of the earth's surface. Examples: forest, savannah, etc. Habitat: habitat is defined as a set of resources (food, shelter) and environmental conditions (biotic and abiotic) that determine the presence, monitoring and reproduction of a population (Gaillard *et al.;* 2009).

Extinction: the disappearance of a species on earth or in a geographical region according to (Malcolm and James, 2007).

Geographic Information System (GIS): an information system that lets you create, view, search and analyze spatial data.

Remote sensing: a technique for capturing and recording electromagnetic energy waves emitted or reflected by the earth's surface.

Georeferencing: georeferencing involves using geographic coordinates to assign a spatial location to map features.

I-2. Contextual framework

I-2-1. Land-use change and biodiversity loss in tropical forests

Land-use change is the greatest threat to terrestrial biodiversity in the tropics (Sala *et al.;* 2000; Jetz *et al.;* 2007; Pekin and Pijanowski 2012). Tropical forests are the richest terrestrial habitats with 50% of the world's species (Dirzo and Raven 2003; Wright 2005) but 68,000 km^2 of tropical forests are lost annually (FAO and JRC 2012). Fragmentation is the biggest threat to forest ecosystems (Bierregaard, Jnr. *et al.;* 2001). Fragmentation can occur naturally through fire (Pickett and Thompson, 1978), but it is the most important cause on a large scale, not to mention the expansion of human land use (Burgess and Sharp, 1981). Habitat fragmentation has three components: loss of original habitat, reduction in habitat size and increased isolation of fragmented habitat, all of which contribute to a decrease in biological diversity in the original habitat (Wilcox 1980; Wilcox and Murphy 1985). Habitat degradation and destruction due to anthropogenic actions are major causes of global biodiversity loss (Brooks *et al.;* 2006). In forests, alteration of vegetation structure and habitat fragmentation through deforestation and forest degradation are among the main threats affecting biological diversity (Sekercioglu 2002, Heikkinen *et al.;* 2004, Chace and Walsh 2006). Bird forests are particularly susceptible to alterations in vegetation structure. Forests expand because of their complex social structure and dependence on vertical vegetation structure (Martin and Possingham, 2005; Davies and Asner 2014). Fragmentation negatively affects bird species (Danjuma *et al.;* 2014)

I-2-2 An overview of Babblers in general

The term Timalie (Babbler in English, phyllanthe or Timalie in French) was first used by the British Dr William Serle in 1949 (Birdlife international, 2016) to refer to the large-tongued birds of forests.   Timalies are birds of the class Aves, order passeriformes, family Leiothrichidae (Serle, 1949) or family Timaliidae (Collar and Stuart 1985).

In 1998, the Bakossi area, one of the most important sites for the species in Cameroon, was teeming with several hundred individuals, as a preliminary population, the species estimated to number around 10,000-19,999 individuals of which 6,667-13,333 are mature (Birdlife international 2014). However, out of 10852 recorded species of all statuses, the order Passeriformes totals 132 families and 6454 species while the family Leiothrichidae has 135 species (Gill *et al.;* 2013). Currently, the diversity of Cameroon's avifauna is estimated at 928 bird species in total, of which 11 are endemic and 8 are rare or accidental (Birdlife international, 2014). However, the White-throated Babbler is a sedentary bird that has been observed in few localities in western Cameroon and eastern Nigeria (Collar and Stuart, 1985). It is a mountain forest bird that was first collected on Mount Kupe and the Obudu Plateau at an altitude of 1520m (Collar and Stuart, 1985). Numbers are not known, but it is generally found in primary forest canopies (Collar and Stuart 1985). The species belongs to the order Passeriformes, family Timalidae and subfamily Timalinae (Collar and Stuart, 1985). The Timalies, family Timalidae, is an important group of insectivorous passerines in the Old World (Sibley and Monroe 1990). There are 257 species of Timalies worldwide, all incorporating 11 genera, distributed for the most part in Central Eurasia, Eastern Eurasia, Africa, Madagascar, the Philippines, Eastern India and Australia (Austin 1987), Cameroon and Nigeria in the case of the White-throated Timalie.

The Leiothrichidae family comprises 4 genera: Blackcap babbler (Turdoites reinwardtii), Brown babbler (Turdoites plebejus), Capuchin babbler (Phyllanthus atripennis) and White-throated mountain babbler (Kupeornis gilberti) according to Cibois (2003). The Kupeornis genus comprises 3 species: Chapin's babbler(*Kupeornis chapini*), Red_collared babbler (*Kupeornis rufocinctus*) and White-throated mountain babbler(*Kupeornis gilberti*) according to (Collar and Robson 2007).

I-3 Knowledge of the White-throated Babbler in Cameroon

I-3-1 Classification according to Serle 1949

Kingdom: Animal

Phylum: Chordata

Class: Aves

Order: Passeriformes

Family: Leiothrichidae

Genre: Kupeornis

Species: *Kupeornis gilberti Serle*, 1949, Kupe Mountain, Cameroon.

The White-throated Babbler is a globally threatened bird species with an endangered status (Barrow and Demey, 2004). The species belongs to the order Passeriformes, family Timaliidae and subfamily Timalinae (Collar and Stuart, 1985). It has only been recorded from a few localities in western Cameroon and eastern Nigeria (Collar and Stuart, 1985). The specimen type was first collected in 1948 at an altitude of 1,520 m on Mount Kupe (Collar and Stuart, 1985).

Figure 1: White-throated Babbler: Kupeornis gilberti (Marcel, 2012)

I-3-2 Description and morphology

The White-throated Babbler was first described by the British Dr. Serle William in 1949 at Mount Kupe (Serle, 1949). These birds are called babblers because of their habit of singing while making noise in small groups "chak", "chook" or "chrook" (Collar, N. & Robson, C. 2017; Meeta kumari 2013). Very little information exists on this species. Timalies, family Timaliidae, an important group of insectivorous passerines of the Old World (Sibley and Monroe 1990). Passerines belong to the class Aves, Order Passeriformes and family Timaliidae. Babblers make up more than half of all bird species. However, the White-throated Babbler is generally found in primary forest canopies, mainly insectivorous, foraging in moss, epiphytes and crevices (Collar and Stuart, 1985).

White-throated Babbler nests are built in trees, in wooded areas or on ledge rocks (Walters 1994). During the breeding season, White-throated Babblers strongly prefer their nests built on surfaces where **they** can hide well from their enemies. Nests are frequently made of lichens, structural leaves and even spider webs.

Morphological characteristics of the Timalies group are generally the absence of distinct juvenile plumage, shape and size, particularly legs and beak as well as sexes are the same in many species (Cibois *et al.;* 2003). They have round wings, long tails, strong legs and beaks (Meeta kumari 2013). Timalies range in size from 21-23cm and weigh 64g, while their colors vary from plain brown and gray with the face and breast white. A remarkable trait of these non-migratory insectivorous birds is a degree of sociability that manifests itself in different ways (Collar, N. & Robson, C. 2017; Meeta kumari 2013).

I-3-3 Geographical distribution

The White-throated Babbler is a species restricted to a few localities in Western Cameroon (Rumpi Hills, Bakossi Mountains, Banyang Mbo Wildlife Sanctuary (R. Fotso in litt.1999), Mont Kupe, Mont Manenguba (Dowsett-Lemaire and Dowsett 1999c), Mont Nlonako, Foto near Dschang and Eastern Nigeria (Plateau Obudu). However, the two most important sites are the Bakossi and Rumpi Hills, where several hundred individuals were estimated (Dowsett-Lemaire and Dowsett 1998d). The population of this species seems to be in severe decline due to the destruction of its habitat, which is why it is classified as threatened (Birdlife international, 2016).

I-3-4 Ecology of the White-throated Babbler

Knowledge of the habitat, distribution, ecology, diet and population dynamics are all important factors in the rational management of a wildlife resource. However, little information exists on the Timalie, as it has not yet been studied in depth. Fruits, insects (or other small black invertebrate animals) and certain seeds are the foods generally consumed by Timalies. Most of the insects eaten by Timalies are plagues that consume common crops or saved seeds (Dhindsa *et al.;* 2000). It has been recorded that its breeding season in West Cameroon runs from June to November, and on the Obudu plateau in April. It appears to be dependent on primary mountain forests with high rainfall, but has also been seen in mature secondary forests at Mt Manenguba (Dowsett-Lemaire and Dowsett 1999c). However, its distribution is linked to the presence of resources, and it has been found at altitudes between 950-2130 m.

I-3-5 Threats and conservation status

11% of birds are threatened with extinction due to human activities, according to the latest report from the Conference of the Parties to the Convention on Biological Diversity (CBD). These human activities are responsible for the decline in avifauna, including unsustainable agriculture, illegal trade, deforestation, dam construction and evasive species, according to the 13[e] summit of the organization held in Cancun, Mexico. According to the IUCN, 11% of the 72 newly recognized bird species are threatened with extinction due to illegal trade, agriculture, logging or invasive species. All these factors have earned it the status of threatened species B1ab(i,ii,iii,v) ver 3.1 according to the IUCN Red List categories since 2016 (IUCN 2014, Birdlife international 2016),i.e. the reduction in its numbers is greater than or equal to 70% estimated over the last 10 years or 3 generations and according to criteria such as: (1) population severely fragmented or present in no more than five localities; (2) continuous decline, observed, inferred or predicted from one of the elements(area of occurrence, area of occupancy, area, extent and/or quality of habitat, number of localities or sub-population, and number of mature individuals); (3)population estimated at 2,500 mature individuals with at least 95% of mature individuals united in a sub-population. According to the 1994 Forest and Fauna Law (MINEF, 1994) and its implementing decree, this is a class A species, i.e. a fully protected species that may not be felled under any circumstances. However, its capture or possession is subject to an authorization issued by the administration in charge of wildlife (MINEF, 1994).

CHAPTER III: MATERIALS AND METHODS

II.1 Study area

This study was carried out in the Bakossi National Park in southwest Cameroon.

II.1.1. Location and context of Bakossi National Park

Bakossi National Park was created by decree N° 2007/1459/PM dated November 28, 2007. With a total surface area of 29,320 hectares, the park lies between 4 50°N and 5 20°N latitude, and between 9 30° and 9 46°E longitude, in the South-West Cameroon region and in the Kupe-Manenguba department.

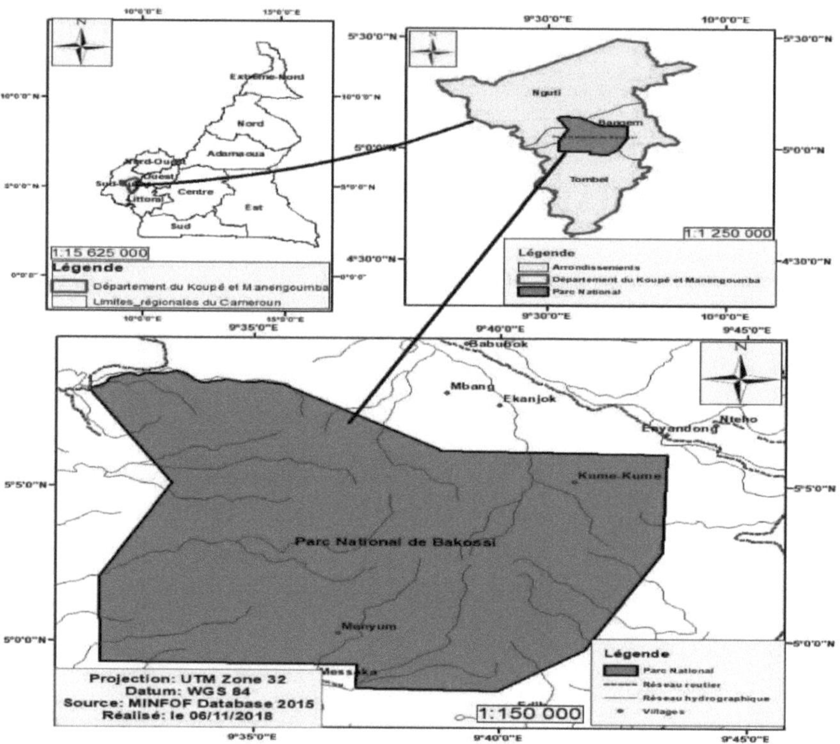

Figure 2: **Location of study area**

II.1.2 Data collection

II.1.2.1 Downloading satellite images

Landsat 7 and 8 images with 30m resolution were uploaded to Earth explorer in February 2019. The reality of image availability and quality led us to adopt images from 2006 and 2018, since no images from 2007 (when the park was created) were available online. These two satellite images were downloaded during the dry season for better separation of vegetation classes and more favorable atmospheric conditions. These images were also chosen because they had the same properties: same spatial resolution, superimposable on the same dimensions and almost the same number of canopy classes, respectively (see figure 3 and 4). Analysis of the Landsat images acquired and processed in 2006 and 2018 was adopted to identify the different forms of land cover in Bakossi National Park. In order to corroborate the information obtained from the maps, GPS coordinates were collected during fieldwork for surveys of vegetation, human activities (agricultural plantations) and bare soil. The GPS coordinates were exported as a table in the Excel spreadsheet. Three types of vegetation formation were taken into account in this study: **primary and** secondary forests, grassy strata and dense forests. The data collected was transferred to a specialized Geographic Information System (GIS) database.

The spectral bands used were red (Channel 3, sensitive to leaf chlorophyll absorption), near infrared (Channel 4, sensitive to leaf structure) and mid-infrared (Channel 5, sensitive to leaf water). These data were used to produce the land cover maps. To ensure the best representativeness, classified images were visualized in Google earth.

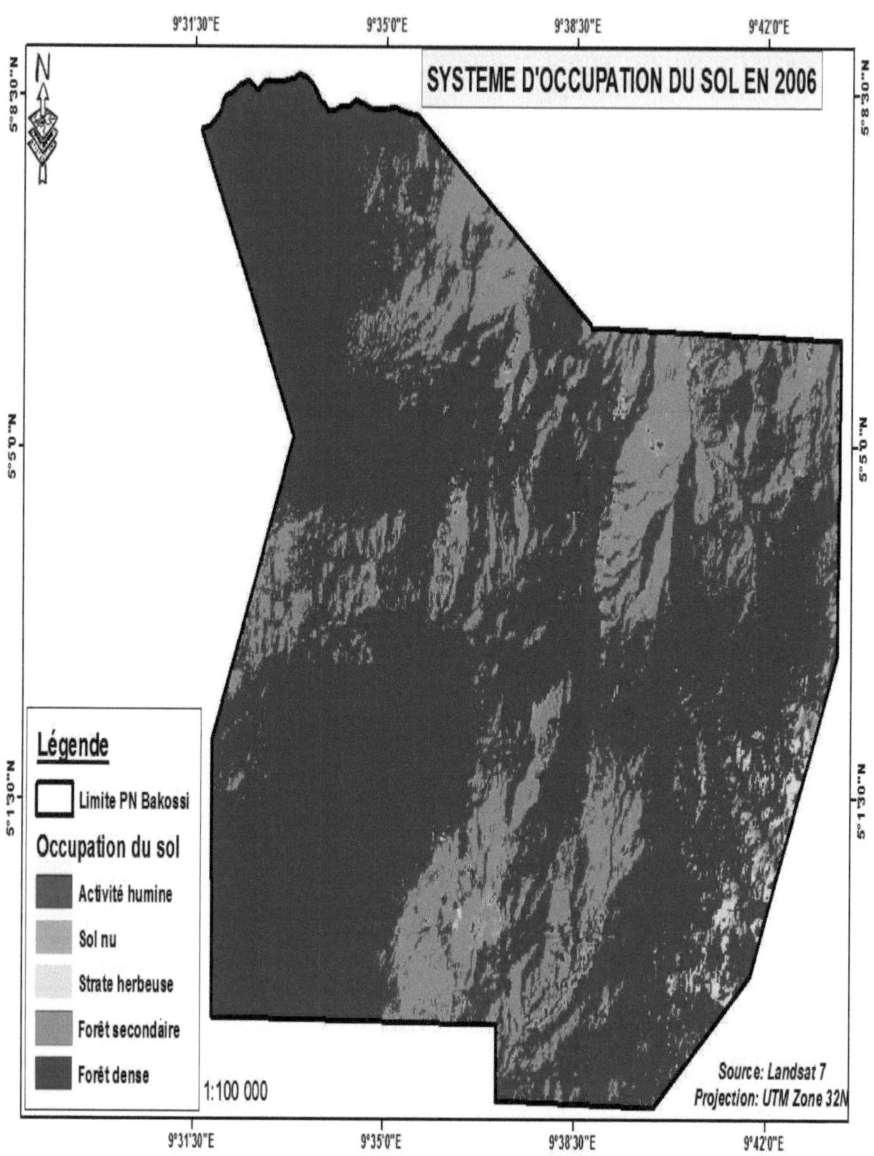

Figure 3: Map of the study area in 2006

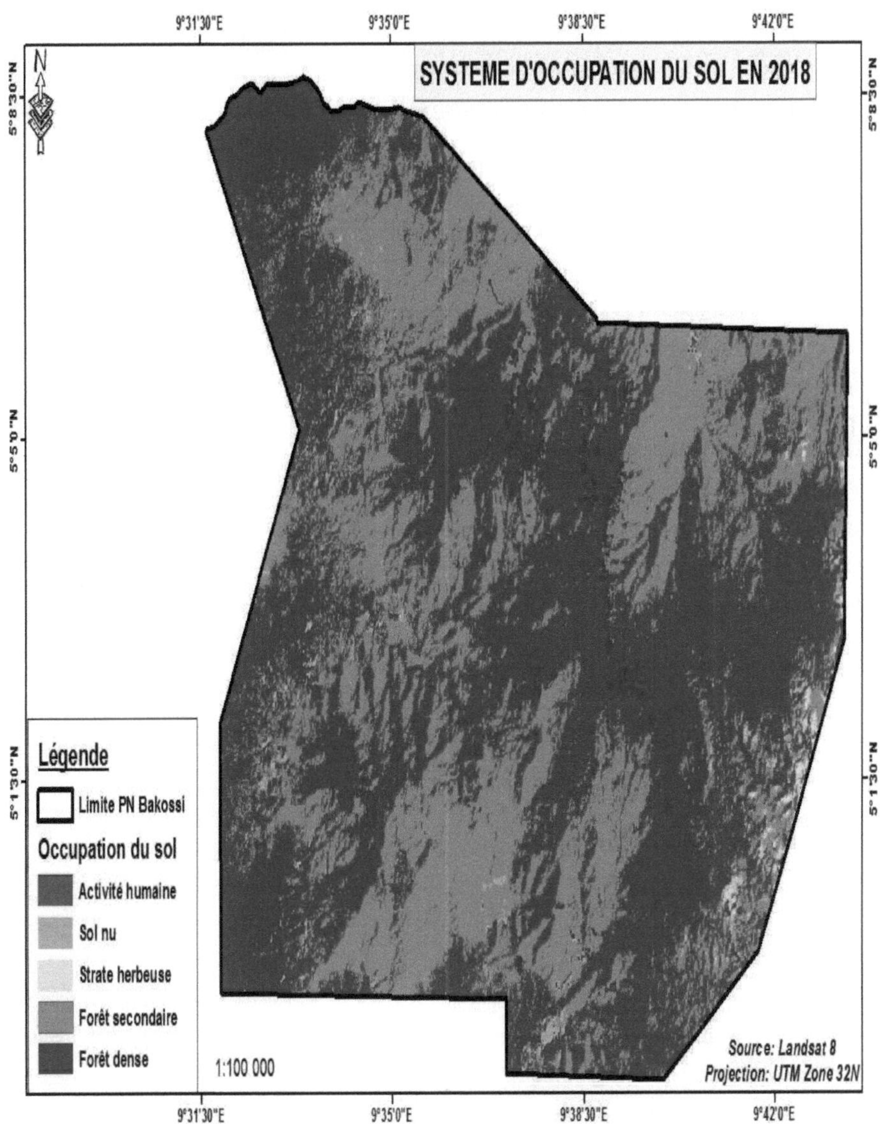

Figure 4: Map of the study area in 2018

These two landsat satellite images explain the land use system between 2006 and 2018 in Bakossi National Park. Through these images, we can distinguish:

- Human activities include agricultural plantations such as coffee, cocoa and banana plantations, as well as logging, houses and bush fires in the park.
- Soil is any surface that has lost all its vegetation, due to agricultural practices.
- The grassy stratum is vegetation with a layer of grass and small trees less than 5 m high.
- Secondary forest is an area with large trees but with some traces of human activity, such as wood cutting and bush fires.
- Dense forest is vegetation with large trees showing no signs of human activity.

II.1.2.2 Field reconnaissance walk

This stage enabled us to gather information to validate the phenomena observed on the landsat images. During this reconnaissance walk, we stopped to verify the presence of the species and count the number of individuals in the various habitat types. This phase also enabled us to gather information on the agents responsible for the degradation of the study area.

II.1.2.3 Questionnaire stage

Interviews were conducted with the local community to find out about their perception of the species, the threats it faces and the driving forces behind the degradation of the area (see questionnaire in appendix).

II.1.3 Supervised classification

The polynomial method of first-order approximation was used to process the LANDSAT images. In fact, it was important that these images were processed in compliance with canopy category classification criteria. All the satellite images obtained were geo-referenced in the UTM coordinate system, which had been used to register these Raster images in order to give them the same geometric properties. Ignoring the edges of the images, pixels (or the smallest unit of an image) with a radius value equal to zero were eliminated. We then proceeded by simple physical recognition (by visualization on a computer screen) to distinguish the different cover classes (dense forest, secondary forest and the grassy stratum, bare soil and human activities). In reality, we proceeded first by unsupervised classification and then by supervised classification using field knowledge to validate phenomena that were not clearly perceptible in the images. The images acquired were processed using ENVI software and the maps were produced using ARCGIS 10.4 cartographic software.

The classified images were transferred to Arc GIS 10, where the varieties of the cover classes were digitized and the statistics recorded on an Excel table, which would eventually help in calculating the surface area of each cover class.

II.1.4 Data analysis

> **Satellite image processing**

Once the Landsat 7 and 8 images had been downloaded, ENVI software was used for image processing and area calculation. A land cover analysis was carried out using NDVI calculation and supervised classification by maximum likelihood. NDVI was calculated using formula 4 (Rouse *et al.;* 1974).

$$NDVI = (Pir - VIS) /(Pir + VIS) \ (1)$$

Where NDVI is the Normalized Difference Vegetation Index, Pir is the near-infrared spectral band and VIS is the red spectral band of the sensors considered. We then visualized the different spectral signatures and performed supervised classification using maximum likelihood.

- **Calculating changes in land use types between 2006 and 2018**

The evolution of land use types was calculated by calculating the areas lost or gained during the study periods. This evolution was evaluated by the difference between the area of a land cover type in year two (2018) and in the reference year (2006). To assess canopy change between 2006 and 2018, Otukei's (2006) formula was used: **Percentage of class area=** AC/ATx100; ΔC=%A_2 (recent year) - %A_1 (old year); TC (%)=A1-A2/A1x 100; **TVA**= ΔC/ ΔP; where **AC**= canopy class area; **AT**= total area; Δ**C**= canopy change, and **AP**= time difference. VAT/ annual rate of change, Δ**A**: rate of canopy change and Δ**T** is the time difference where ΔT= t -t_{21} .

> **About the species**

The data collected in the field were entered and saved in Microsoft Excel 2010 for the various statistical tests. Descriptive analyses of the data were carried out using Microsoft Excel 2010. Descriptive statistics enabled us to estimate the probability and means of encounters in the different habitat types. They also showed how drivers and agents affect the species' habitat.

II.2.1. Species inventory

During the reconnaissance walk in the field, we looked for the presence **or absence of** the species, counted the number of individuals at each station and determined the type of habitat where the species had been seen. Data was collected along the altitudinal gradient.

- Inventory **techniques**

Three methods were used to obtain the results we were looking for. The first was remote sensing coupled with GIS, which enabled us to determine the evolutionary trend of the vegetation cover in Bakossi National Park. These methods were used because they offer several advantages: they are less time-consuming, less costly and quicker to use, and they enable us to assess changes in vegetation cover over large areas (Gottschalk *et al.;* 2004; Loveland and Dowyer, 2012; Hansen *et al.;* 2013). The second method consisted in the administration of a questionnaire to gather information on the factors responsible for the change in vegetation cover, and finally the field reconnaissance walk to identify the different types of habitat used by the White-throated Babbler in order to deduce the most frequented habitat.

Remote sensing, GIS and questionnaire techniques were used in the inventory of the White-throated Babbler population in Bakossi National Park. Remote sensing and GIS are the most widely used techniques because they reduce time wasted, are less costly and quicker to carry out, and enable the assessment of vegetation cover change over large areas (Gottschalk *et al.;* 2004; Loveland and Dowyer, 2012; Hansen *et al.;* 2013).

- **Inventory process**

Once the satellite images had been obtained, a raid was carried out in the field to confirm the realities observed on the maps. As remote sensing and GIS did not provide information on the species and its habitat, we met with the village chiefs to obtain permission to survey the population. With the collaboration of the chiefs or their representatives, we were given a **guide** in each village, who directed the survey in relation to the identification of the respondent. A questionnaire was administered individually to the local population to gather the following information: knowledge of the species, habitat frequented, age, sex, season of encounter, occupation of the respondent, level of education, number of plots of land, form of agriculture, reasons for hunting for hunters and finally the usefulness of the species. The aim of collecting all this information was to find out whether the local population knew about the species and its habitat. During the period from December 2017 to January 2018, we carried out a field walk to look for the presence of the species by habitat type. We would wait for 3 minutes to follow the response(s) of individuals of the species around the station according to Danjuma *et al.;* (2014). The following information was collected: geographical coordinates (longitude, latitude and altitude) of each point, altitude, presence of the species, habitat type and crop types when plantation fields were involved. Data collection took place daily between 7 and 11:30 a.m. and in the evening between 3:30 and 5:30 p.m., as birds are most active at these times (Bibby *et al.;* 2000; Dennis et al 2002).

Figure 5: Control image with Deck population

II.2.2. Anthropogenic factors and the different habitat types of the White-throated Babbler.

II.2.2.1. Habitat types of the White-throated Babbler.

The aim of this project was to study the different habitat types of the White-throated Babbler in order to determine the habitat it uses most in Bakossi National Park. The various habitat types were determined by visual assessment, given that quantifying environmental characteristics is very complex and time-consuming, according to Bibbly *et al.* (2000). The different types of habitat identified in the field are:

> ➢ Primary or dense forest: this is a forest with no trace of human activity, characterized by the presence of the crowns (**tops of trees**) of tall trees over 20 m high, the upper canopy (**edge of a field or tree**) is the forest canopy, the presence of lianas and the presence of creeping and climbing plants.

- ➢ Secondary forest is characterized by the presence of trees grouped into two strata: the low tree stratum, with trees between 7 and 15 m high, and the high tree stratum, with trees over 15 m high. In this category of forest, we note the presence of human traces such as wood cutting and bush fires.
- ➢ Herbaceous savannah is vegetation made up of small plants no taller than 1m (grasses, ferns, under-shrubs and young shoots). The presence of bush fires is often a sign of human activity in this habitat.
- ➢ The shrub savannah is vegetation made up of small plants no more than 1 m high. There are also a few small trees less than 5 m high.
- ➢ Cultivated fields are spaces used for agricultural practices.

II.2.2.2. Anthropogenic activities

Observations of signs of human presence took place at the same time as the census of the different types of habitat frequented by the White-throated Babbler. During the reconnaissance walk, the aim was to record all signs of human presence at the various stations sampled. The following clues were identified: wood harvesting (whole trees or firewood), animal traps, cartridge casings, presence of hunters' shelters/repaires, presence of bush fires and burning periods, presence or absence of fields, type of crop (perennial, annual, polyculture or monoculture or unknown). On arrival at each collection station, the various activities were recorded by direct observation in the field and recorded on data collection sheets. In the questionnaire administered to the local population, we asked whether the forest area had decreased or increased. If so, what were the factors responsible for this reduction in area: agriculture, bush fires, logging, population growth, use of wood for building houses, etc. (See questionnaire sheet on the appendix page).

III.3 Abiotic characteristics of the study area

III.3.1 Climate

The Bakossi area, located in the South-West region, is dominated by a tropical climate. It is characterized by two distinct seasons: the dry season, from November to March, and the rainy season, from April to October. Rainfall is abundant, with an annual average of around 3,000mm. The maximum temperature is 30°C, while the minimum is 10°C on the slopes of Mt Kupe. Thanks to this climate, Bakossi's land is dominated by dense equatorial forest (Denise and Milena 2017).

III.3.2. Geomorphology and pedology

Bakossi soils are naturally rich. However, these soils located on the slopes of Mont Kupé and Mont Manenguba are of volcanic origin and are highly fertile (Miavita 2011). The fertility of the soil in this area is an asset for the population in the practice of two of Africa's best-known forms of agriculture: subsistence farming (maize, potatoes, manioc, yams etc.) and cash or commercial farming (coffee, cocoa and oil palm). However, although the area is dominated by fertile soils, some places are interrupted with sand, **clay, loam (greasy and impermeable clay)** others are sedimentary soils that are also suitable for production cultivation (Denise and Milena 2017).

III.3.3. Hydrographic networks

The Bakossi area is surrounded by high-altitude mountains ranging from 200 to 2396m above sea or ground level. These mountains are: Mount Kupe (2,050m) and Mount Manenguba (2,396m). According to Miavita (2011), the Bakossi region is established on the fault line that originates in the Atlantic at the peak of Sao Tomé, with a south-west - north - east orientation. This fault line passes through Monts Bamboutos, Manenguba, Kupe and Mont Bakossi, continues along the coast where it ends in the Tibesti in Chad, so these mountains are all of volcanic origin. The slopes of these mountains are made up mainly of fertile volcanic soils that are suitable for cultivation and a large, dense rainforest rich in both flora and fauna. Generally, Bakossi's land is of low gradient, punctuated with few ridges and hills whose valleys have been deepened by runoff. Despite abundant rainfall, these two arrondissements are drained by few rivers, with the Mungo River and Bemweh Lakes highly significant as tourist attractions (Denise and Milena, 2017).

III.4. Bio-ecological characteristics

III.4.1 Vegetation and flora

Bakossi is dominated by dense equatorial forest in the arrondissements of Tombel, Nguti and Bangem. Forests can be found almost everywhere on the slopes of Mont Kupe and on the plains of Bangem. The vegetation here is mainly savannah, extending as far as the Melong area in the coastal region. This vegetation is used by the Bororo for cattle breeding. Vegetation varies with altitude. Sub-mountain forests extend from 900 to 1800m altitude. Typical afromontane species are: ***Nuxia congesta, Podocarpus atifolius, Prunus africana, Rapanea melanophloeos and Syzgium guineense bamendae.***

III.4.2. Fauna

The Bakossi forests are a suitable habitat for many animals and consequently, as in most dense equatorial forests, they are rich in wildlife. However, the Bakossi forests are an important site in Cameroon for certain globally threatened species, including seven strictly endemic bird species: Bamenda apalis (*Apalis bamendae*), Bangwa forest warbler (*Bradypterus bangwaensis*), white-throated mountain-babbler(*Kupeornis gilberti*), banded wattle-eye (*Platysteira laticincta*), Bannerman's weaver (*Ploceus bannermani*), Mount Kupe bush-shrike (*Telophorus kupeensis*) and Bannerman's turaco (*Tauraco bannermani*). The latter is a cultural icon for the Kom people who live in this area. Eleven species of small mammals are endemic to this area: Eisentraut's striped mouse (*Hybomys eisentrauti*), the Mount Oku hylomyscus (*Hylomyscus grandis*), Mount Oku rat (*Lamottemys okuensis*), Mittendorf's striped grass mouse (*Lemniscomys mittendorfi*), Dieterlen's brush-furred mouse (*Lophuromys dieterleni*) and Eisentraut's brush-furred rat (*L. eisentrauti*), Oku mouse shrew (*Myosorex okuensis,*) Rumpi mouse shrew (*M. rumpii*), western vlei rat (*Otomys occidentalis*), Hartwig's soft-furred mouse (*Praomys hartwigi*), and Bioko forest shrew (*Sylvisorex isabellae*). Several primates are threatened with extinction in this region, including: Cross River gorilla (*Gorilla gorilla diehli*), a subspecies of western gorilla: mainland drill (*Mandrillus leucophaeus leucophaeus*), Preuss's red colobus (*Pilocolobus preussi*), common chimpanzee (*Pan troglodytes*) and Preuss's monkey (*Cercopithecus preussi*).

III.4.3. Population and socio-economic activities

In 1976, the population of the Sud-Ouest was estimated at 620,515; by 2015, it had grown to 1,534,232 (BUCREP, 2010b). In 2010, over 52% of the Sud-Ouest's population lived in rural areas (BUCREP, 2010a). Today, more than 250,000 people live in Bakossi. This region is overpopulated by young people from the North-West, West Cameroon and neighbouring countries such as Nigeria, with significant numbers looking for work on large plantations such as the CDC (Cameroon Development Corporation), at the Limbe petrol refinery, or to create new farms on the fertile soils (a phenomenon explained by the uncontrolled occupation of land in this area) according to Lairds *et al.;* (2007). The main occupation of the inhabitants is farming, as many take advantage of the fertile soils and good climate (Denise and Milena, 2017). Around 80% of the population is involved in agriculture and agriculture-related activities. The Sud-Ouest region includes population, housing, industrial plantations, an oil refinery complex, extensive logging by companies and many other industries, equipment and infrastructure (Zogning *et al.;* 2010). Livestock farming is very little developed in the locality, being extensive and characterized by small herds (**all the livestock on a farm or in a region**), with livestock **wandering.** The main species raised include poultry (chickens), sheep, goats and pigs. This breeding is oriented towards both consumption (poultry) and marketing (goats). (CVUC, 2015). Taken together, these activities constitute sources of income to ensure the survival of the population.

III.5. Study sites

Studies carried out on the Obudu Plateau in Nigeria and Mount Kupe in Cameroon reveal that the White-throated Babbler is a forest specialist, frequently found in primary forest canopies and rarely in secondary forests (Collar and Stuart 1985). The Bakossi area is cited as one of the species' most abundant sites, with a high population density in contrast to other sites (Dowsett-Lemaire and Dowsett 1998d). The White-throated Babbler is listed as globally threatened according to Borrow and Demy 2004, but its distribution remains limited in this area due to the presence of the Bakossi forests. In view of previous work on avifauna, which was limited to inventories, this study carried out in the GNP aims to determine the impact of habitat deterioration and the effects of human activities, given the decline in its population.

CHAPTER IV: RESULTS AND DISCUSSION

IV.1 RESULTS

IV.1.1 Change in cover type at Bakossi National Park

The canopy types obtained from on-screen scanning after supervised classification are presented in **Table i (4)** following the statistical values acquired.

Table i: Summary of changes in habitat types at PNB between 2006 and 2018.

Canopy type	A_1 2006 (ha)	% A_1	A_2 2018(ha)	% A_2	ΔA (%)	TC (ha)	TVA
Dense forest	23283.36	79.41	19472.31	66.41	-13	-16.37	-1.083
Secondary forest	5244.12	17.89	9138.78	31.17	13.28	74.27	1.106
Savannah	329.58	1.12	227.16	0.77	-0.35	-31.07	-0.029
Bare ground	338.85	1.16	240.6	0.83	-0.33	-28.98	-0.027
Human activities	124.38	0.42	241.38	0.82	0.40	0.0094	0.033
Total	29320.29	100	29320.29	100			

A_1 **2006 (ha)**: area of canopy habitat classes in 2006 in hectares; A_2 **2018(ha)**: area of habitat classes in 20018 in hectares; % A_1: percentage of habitat classes in 2006; % A_2 : percentage of habitat classes in 2018; ΔA **(%)**: change in habitat classes in percentage; **TC (ha)**: rate of change in land cover between 2006 and 2018 in ha; **K: rate of regression or expansion**; **TVA**: rate of annual change.

Table **(i)** 4 shows that in 2006, the surface area of habitat classes in the GNP was respectively 79.41% for dense forests; 17.89% for secondary forests; 1.12% for savannah; 1.16% for bare soil and 0.42% for human activities. Nevertheless, in 2018, the percentage of surface area of each type of cover was 66.41%; 31.17%; 0.77%; 0.83% and 0.82% respectively, in the same order as above. It is clear that changes have taken place in the different cover classes between 2006 and 2018 in the light of these data (Figure 2).

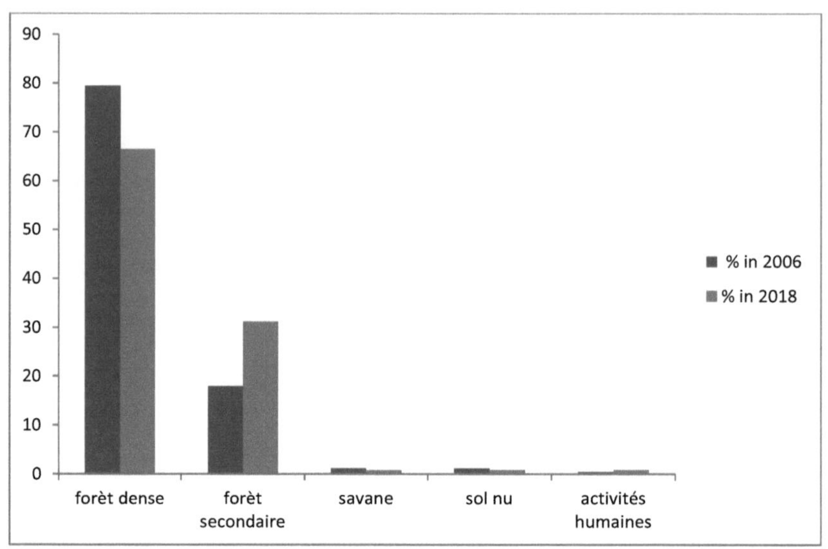

Figure 6: Overall variation in habitat classes observed at PNB between 2006 and 201 8.

In-depth analysis shows that between 2006 and 2018, dense forests, savannah and bare soil respectively recorded a loss of surface area of -13%; -0.35% and -0.33%, yet during the same period, secondary forests and human activities increased by 13.28% and 0.4% respectively.

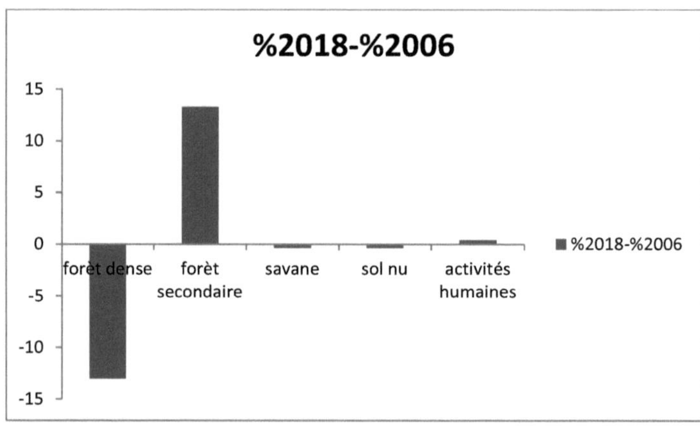

Figure 7: Annual variation in habitat classes observed at PNB between 2006 and 2018.

This change in habitat classes between 2006 and 2018 is clearly visible (habitat class variation figure), and manifests itself in either an increase or decrease in the area of each habitat class (figure 10).

IV.1.2 Agents responsible for habitat class degradation

During the field reconnaissance walk, several signs of threats to white-throated blackbirds and biodiversity were noted in the GNP. These included hunting (cartridges, non-selective traps and hunter's huts, which were much more common in the villages of Kodmin, Muahumzum and Zimbeng), habitat degradation (bush fires and illegal logging: these were noted in Kodmin (see appendix) and agricultural activities within the GNP (these were noted in the village of Deck, where only secondary forest and coffee and cocoa plantations predominate). Pressure on the land is reflected in an increase in agricultural mosaics and industrial plantations. The most worrying type of cultivation is coffee, where large areas of land are destroyed to create coffee plantations (this has been noted several times within the GNP). The growth of the population living in the GNP will increase the need for land for farming, which is all the more worrying as the creation of plantations is responsible for the transformation of es forests into fields for cultivation.

Table ii: Respondents' answers on factors responsible for changes in forest cover

Factors	Number of responses	% total
Agriculture	41	
Bushfire	5	8
Population growth	13	20
House building	2	3
Wood cutting	4	6
Total	**65**	**100**

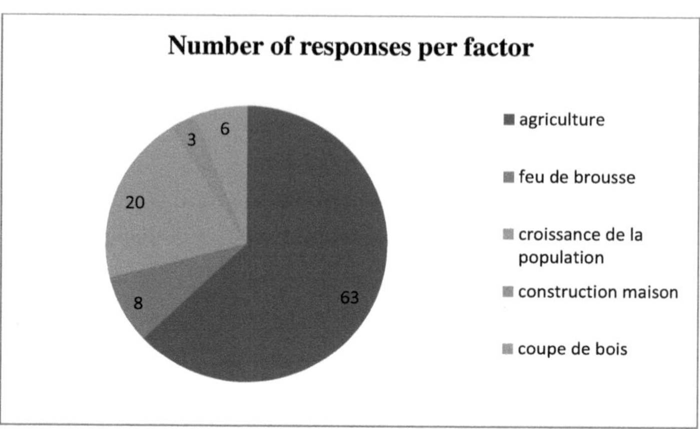

Figure 8: Diagram of respondents' responses to the factors responsible for forest cover change.

The diagram shows that the main factor responsible for the degradation of White-throated Babbler habitat is agriculture (63%). After agriculture comes population growth (20%).

Figure 9: Images of anthropogenic evidence in Bakossi National Park.

A: bird captured by trap; B: logging (timber); C: shell + cartridge; D: road: an example of habitat fragmentation; E: coffee plantation; F: hunter's hut; G: house used by farmers; H: captured animals (monkeys + antelope).

IV.1.3 Information on the White-throated Babbler

During the field reconnaissance walk, we counted 152 individuals in dense and secondary forests. During the inventory, the species was observed 30 times at the sampled points. In the primary forest, the species was observed 25 times and 114 individuals were counted, while in the secondary forest, we had 5 responses for 38 individuals observed, giving a total of 152 individuals counted at all points visited. Apart from these two habitats, the species was not observed elsewhere (grassy savannah, shrubby savannah and bare ground). This result corroborates that of the questionnaire, where 31 of the 65 individuals questioned all answered that the species lives in the forest. This result shows that primary forest is the habitat used by the White-throated Babbler.

Table iii: Summary of White-throated Babbler responses to different habitat types

Habitat type	Number of stations
Dense forest	25
Secondary forest	5
Shrubby sava nnah	0
Grassy savannah	0
Crop fields	0

The inventory took place in five habitat types, as follows:
- ➢ in the points sampled, the species was observed 30 times and 25 times in dense forest, i.e. 83.33%.
- ➢ in the secondary fort, the species was observed five times out of 30, i.e. 16.67%.
- ➢ In the shrub savannah, grassy savannah and crop fields, no species was observed.

IV.1.3.1 Likelihood of encountering the White-throated Babbler

Using the data collected in the field, we calculated the probability of occurrence of the species in each habitat type, the standard deviations and the means of encountering the species in dense and secondary forests (see table iv).

Table iv: Probability of encountering White-throated Babbler

Habitats	Probability		Counting	
	Estimate	Standard deviation	Estimatin	Standard error
Dense forest	0.75	0.0035	2	0.0041
Secondary forest	0.25	0.0035	0.58	0.0008

This table shows that the probability of encountering the White-throated Babbler is high in dense forest (75%) and low in secondary forest (25%).

IV.1.4.2 Respondents' perception of species knowledge

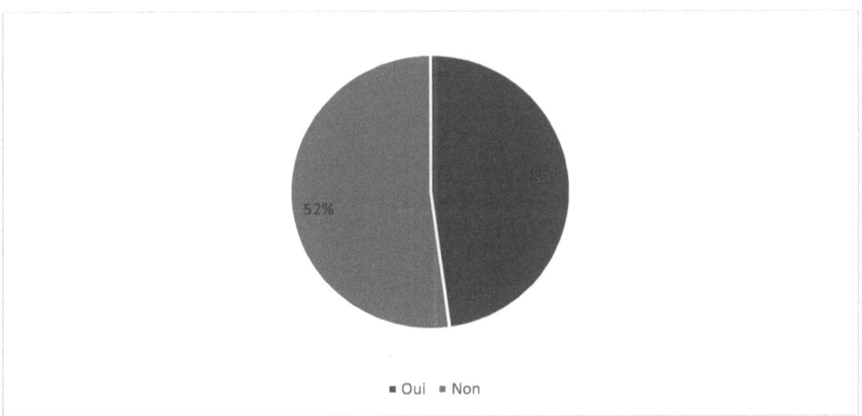

Figure 10: Diagram of respondents' perceptions of knowledge of the species

The figure above shows that 52% of respondents have no knowledge of the species, while some (48%) do have knowledge of the species. The perception of respondents who claim to have no idea about the species for $P<0.05$ (see appendix) is not significantly different from those who claim to have knowledge about the species (Mann-Whitney U Test: $Z= -0.38$; $p= 0.7014$).

IV.1.3.2 Knowledge of species by sex

During the field trip, a questionnaire was administered to 65 individuals, including 42 males and 23 females. At random, 31 individuals had knowledge of the species, i.e. a percentage of 48%, while 34 had no knowledge of the species, i.e. a percentage of 52%. Of the 31 individuals with knowledge of the species, 28 were males (67%) and only 3 females (13%).

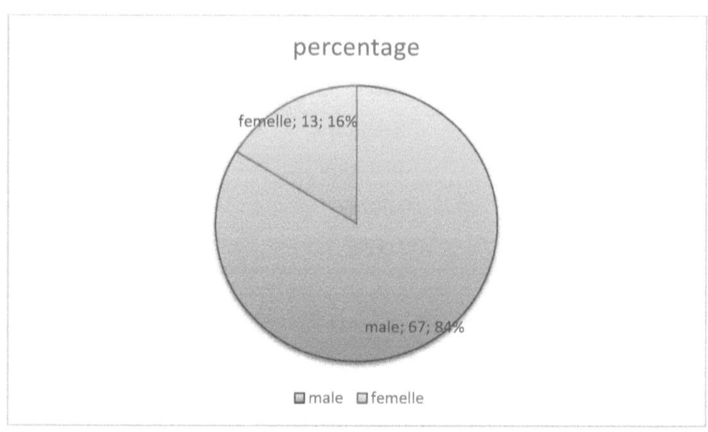

Figure 11: Species knowledge diagram based on sex

IV.1.3.3 Knowledge of the species as a function of age.

During the field trip, a questionnaire was administered to 65 individuals aged between 13 and 70. Between the ages of 13 and 20, only 1 individual was familiar with the species. Between the ages of 21 and 40, 17 individuals were familiar with the species. Finally, between the ages of 41-70, 13 individuals were familiar with the species the species.

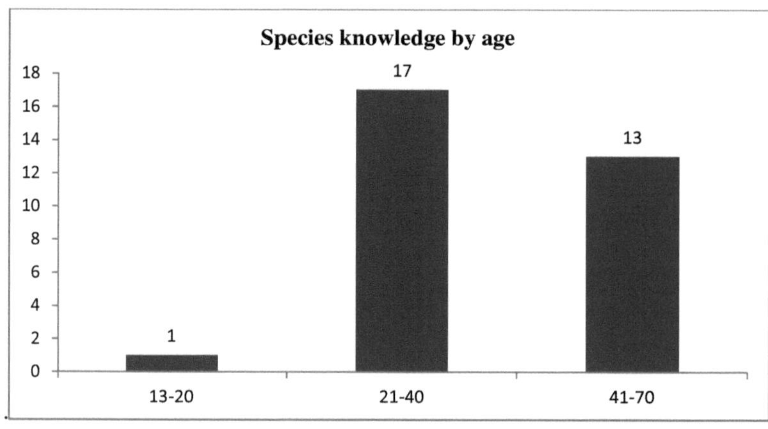

Figure 12: Diagram of species knowledge as a function of age

IV.1.3.4 Respondents' professions

During the field visit, 65 individuals were administered. According to each individual's profession, we have: 14 schoolchildren, 8 housewives, 21 hunters, 15 farmers, 2 shopkeepers and 2 primary school teachers.

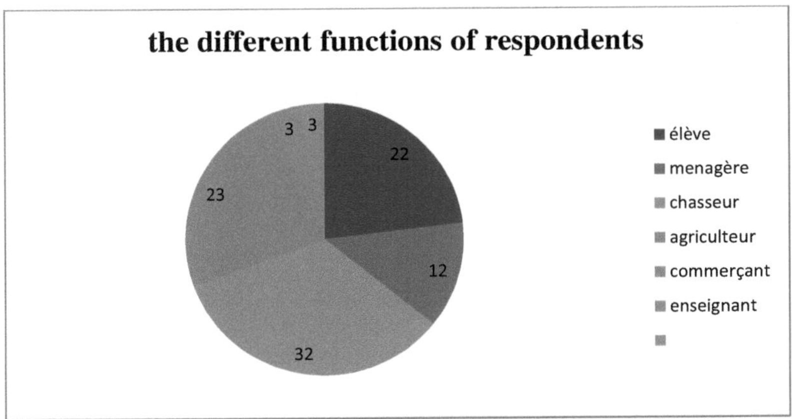

Figure 13: Diagram of the different professions

IV.1.3.5 Species knowledge by profession

During the field trip, a questionnaire was administered to 65 individuals, 31 of whom were familiar with the species. According to profession: 1 student, 3 housewives, 21 hunters, 4 farmers and 2 teachers knew the species. None of the shopkeepers knew the species. We note that all hunters know the species.

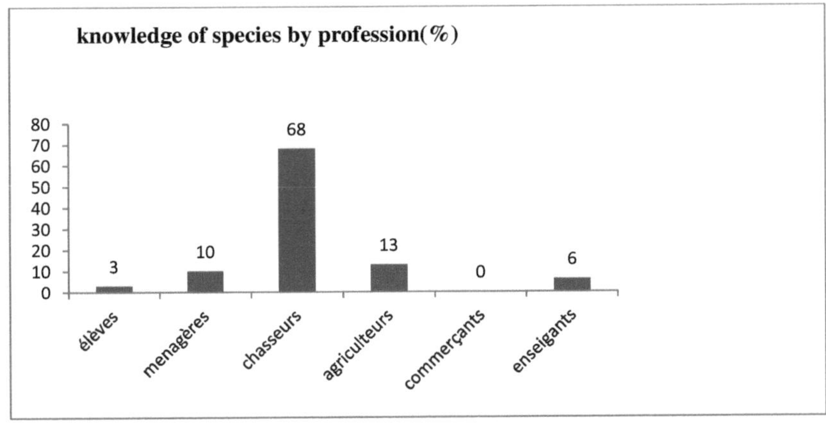

Figure 14: Knowledge diagram by profession

IV.1.3.6 Variation in the number of agricultural plots per group of individuals

With the exception of the students, all the individuals are farmers. Each individual group owns at least two plots: one for growing maize and the other as a coffee or cocoa plantation.

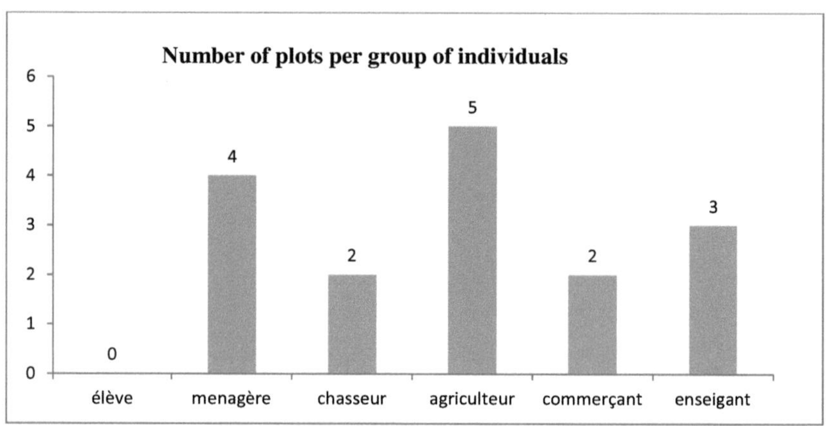

Figure 15: Diagram of the number of plots per group of individuals

IV.1.3.7 Species knowledge by school level

A questionnaire was administered to 65 individuals to find out whether knowledge of the species was a function of educational level. Four levels were identified: primary school without the CEP(a), primary school with the CEP(b), secondary school without the BEPC(c) and secondary school with the BEPC(d). We noted that individuals at primary level were more familiar with the species than those at secondary level. Consequently, knowledge of the species does not depend on the level of schooling.

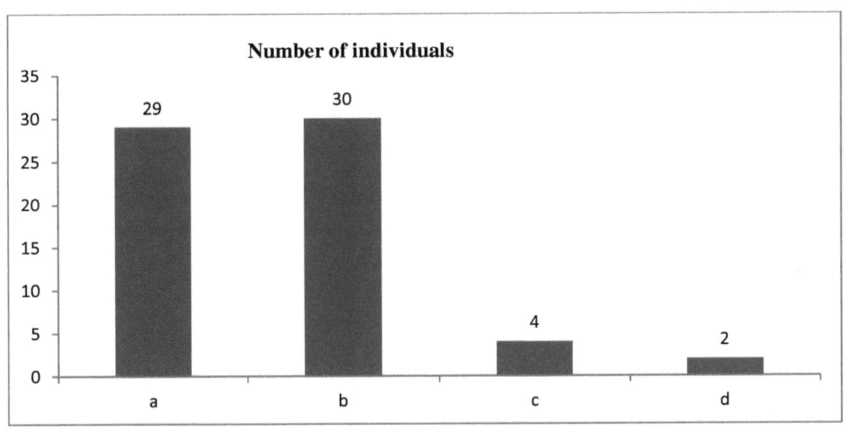

Figure 16: Species knowledge chart by school level

IV.2 DISCUSSIONS

IV.2.1 Canopy change in Bakossi National Park

The remote sensing technique used to assess changes in White-throated Babbler habitats in Bakossi National Park reveals that there has been a decrease in the area of dense forest, savannah and bare ground between 2006 and 2018, while in the same period the areas of secondary forest and human activities have increased. This result confirms that the creation of the GNP has not been followed by the protection of forest cover. This study is similar to those carried out in the Rumpi Hills (North-West) of Cameroon by Mukete *et al;* (2018); in Mount Cameroon in 2016 by Guetse francis, which showed the decrease in dense forest areas in the Rumpil Hill (one of the important sites for the conservation of the White-throated Babbler). In view of the increasing loss of dense forest areas in the distribution sites of the White-throated Babbler in Cameroon, we note a decrease of 13% in 12 years in the Bakossi, i.e. an annual loss of 1.08% per year, 14% in 25 years in the South-West at Lebialem-Mone according to Takem-Mbi and Ngoufo (2015) and also in the Rumpill Hills according to Mukete et al(2018). This area loss is similar to that reported in the literature for West Africa (1%) and Central Africa (0.6%) according to Puig (2001). Given that the White-throated Babbler is a primary forest specialist (Danjuma *et al.;* 2014), and knowing that deforestation has a negative impact on biodiversity (Gibson *et al.;* 2011), we can predict the extinction of the White-throated Babbler in the coming decades if nothing is done to mitigate the conversion of forest cover to cropland. The same observation was made by Takem-Mbi in 2013, when he noted a reduction in forest and an increase in field landscapes in the Bafut-Ngemba Forest Reserve of North-West Cameroon.

In the light of the above observations, we can conclude that human population growth increases the need for resources, hence the exploitation of forest resources, which increases the risk of biodiversity loss. This risk is even higher in regions where the economy is based on agriculture, such as Africa (William, 1900). This study shows that the use of satellite images is an appropriate method for providing rapid information over large areas, on the degree of change in the cover and use of an environment.

Figure 17: Habitat used by the White-throated Babbler.

The literature on the state of conversion of plant cover in the Bakossi reveals that degradation is increasing with the sale of land to foreigners by the natives for the creation of vast palm and cocoa plantations. The CDC is recognized as the largest landowner in the south-west (Lairds et al., 2007; Miavita, 2011). Since the economic crisis of the 1980s in Cameroon, people have become land-independent, according to Schmid Soltau (2003).

IV.2.2 Threat factors observed at the study site

The White-throated Babbler is a globally threatened bird species (Borrow and Demey, 2004). Discussions with GNP authorities, guides and villagers reveal that agriculture is the main threat to the White-throated Babbler (see figure 9), as it contributes to the conversion of forests to agricultural plantations. This is said to be due to the increase in the price of cocoa to 1225 FCFA/kg in 2007 (Elong, 2011). This price increase motivated many individuals to invest in the creation of the coffee and cocoa plantations we found in the park. These plantations were more common in the Deck village. Several authors have asserted that the development of human activities such as agriculture have an immediate impact on land use and a direct impact on forest cover (Mather and Needle , 2000; Geist and Lambin, 2002; Jansen *et al.;* 2008), the same observation was made by Momo solefact in 2018 in Koupa-Matapit in the Noun department. This modification increases the risk of biodiversity loss (William, 1990). Apart from agriculture, we observed: non-selective traps, traces of bush fires, cartridge casings and logging to obtain crop fields and construction timber for dwellings such as households, schools and church. Population growth is estimated to be responsible for 20% of the change in forest cover (see figure 9). This value is well above that obtained in south-west Cameroon at Lebialem-Mone by Takem-Mbi in 2015. In 2010, 52% of the population of South-West Cameroon lived in rural areas (BUCREP, 2010b). This factor increases the need for land to grow crops, and for timber to build wooden houses.

Of the five types of habitat sampled, i.e. dense forest, secondary forest, shrub savannah, grassy savannah and cultivated fields, the rate of occurrence in dense forest was 87% at the 30 stations where the species was observed. 17% on 30 stations for secondary forests This is due to the fact that the White-throated Babbler is a bird specialized in primary forests with closed canopies, but is often observed in secondary forests (Danjuma *et al.;* 2014). This result corroborates with a study carried out in Costa Rica which shows that the forest is the primary habitat of the majority of the limited avifauna and that the birds are highly vulnerable to deforestation (Vicencio Oostra *et al.;* 2008; Birdlife internationnal 2013). This encounter rate is nil in shrub savannahs, grassy savannahs and cultivated fields. It is understandable that the White-throated Babbler is sensitive to habitats altered by human activities.

II.2.3 Threats to the extinction of the White-throated Babbler

Bakossi's growing population is increasing the need for resources, which in turn is leading to changes in forest area due to agriculture. More than 90% of the population lives from this activity, where each individual has an average of three plots of land. Worldwide, 50% of tropical forests have been severely degraded or converted to agriculture (Wright 2005; FAO 2007; Stork *et al.;* 2009). The same observation was made in Oku by Solecfack *et al;* (2012) and in Rumpi Hills by Mukete *et al;* (2018) where a loss of forest area due to conversion to cropland was noted. The main cause of deforestation is the conversion of land into cultivated plots, followed by the felling of trees and finally the passage of fire (Solefack *et al.;* 2012). Many males were familiar with the species, which is justified by the fact that men farm during the day and hunt at night. In contrast to the females, who were unaware of the species, we understand that women are more involved in agriculture, yet arable fields are not a habitat used by the White-throated Babbler.

According to Senapathi *et al. (2007),* changes in the dense forest habitats on which the survival of the White-throated Parrot depends are caused by regular bush fires, which represent a major threat to the species' conservation. Habitat degradation would have a negative impact on the preservation of this species, as observed with grey parrots in Cameroon (Tamungang *et al.,* 2014).

CONCLUSION

At the end of this study, the aim of which was to assess the level of change and the factors responsible for the deterioration of the habitat of the White-throated Babbler, we can say that the remaining population of this species still present in this site is scattered in the primary forest and rarely in the secondary forest. Its probability of occurrence is high in primary forest (75%) and low in secondary forest (25%). This shows that primary forest is the habitat most used by the White-throated Babbler. The presence of the White-throated Babbler was noted overall at 29% in all study stations. The area of dense forest, savannah and bare ground declined between 2006 and 2018 as a result of deforestation and bush fires. During the same period, there was an increase in secondary forest cover and human activities (crop plantations, infrastructure) due to logging. Anthropogenic factors have contributed significantly to the loss of White-throated Babbler habitat. The growth of plantations and deforestation cause the conversion of forest cover to agriculture and the loss of habitat for bird species. Understandably, the creation of Bakossi National Park has not slowed the pace of deforestation. However, to mitigate the threats responsible for the loss of White-throated Sandgrouse habitat, we propose: the delimitation of the GNP, the intensification of patrols and the encouragement of perennial agriculture and, finally, the encouragement of the local population to reforest outside the park.

RECOMMENDATIONS

The results of this study lead us to make the following recommendations:

- ➢ **To the Cameroon government**
  - Raise awareness of the need to combat deforestation among the local population;
  - Organize patrols within the GNP to limit hunting pressure;
  - To limit the GNP in order to deprive the population of access to the park's spaces and resources;
  - Encourage perennial farming for coffee and cocoa outside the park;
  - Raise local awareness of bushfire prevention in savannahs and forests.
- ➢ **To NGOs**
  - Guide the GNP development plan to protect birds
- ➢ **To the public**
  - Avoid slash-and-burn agriculture in savannahs and forests;
  - Stop cutting wood in the park and cut it rationally outside the park instead;
  - Stop farming within the GNP;

OUTLOOK

For the future, we're looking ahead:

- ❖ Reassess the population size of the White-throated Babbler in GNP;
- ❖ Determine the effect of human activities on the distribution of the White-throated Babbler;
- ❖ Study the ecology of the White-throated Babbler to understand its role in ecosystem balance.

REFERENCES

Addisu Asefa, Andrew B. Davies,2b Andrew E. Mckechnie,1 Anouska A. Kinahan, And Berndt J. Van Rensburg2: Effects of anthropogenic disturbance on bird diversity in Ethiopian montane forests: American ornithology:Volume 119, 2017, pp. 416-430

Achard, F., H. Eva, H.-J. Stibig, P. Mayaux, J. Gallego, T. Richards, And J.P. Malingreau (2002). Determination of deforestation rates of the world's humid tropical forests. Science 297: 999-1002.

Analysis: a geographical approach to the protection of biological diversity. Wildlife Monogr. 123: 1-41.

Arnold and Jongma (1997) Firewood and charcoal in developing countries an economic survey. *Unasylva*, 29, 2 - 9

Austin, O. L. 1987. Birds of the world: a guide to the 185 bird families. Twickenham.country life Books.

Bakossi Tribe in Cameroon-BACDA, workong forits empowerment and development by Denise Nanni and Milena Rampoldi, promosaik, 2017.

Bamba, I., Y.S.S. Barima, and J. Bogaert (2010). Influence of population density on the spatial structure of a forest landscape in the Congo Basin, R, D. Congo. Tropical Conservation Science 3, no. 1: 31-44.

Bawa, K.S., And S. Dayanandan. 1997. Socioeconomic factors and tropical deforestation. Nature 386: 562-563.

Bibby, C. J., Burgess, N. D., Hill, D. A., and Mustoe, S. (2000). Bird census biocultural diversity in migrant and indigenous livelihoods around Mount Cameroon. International and cultural diversity in local livelihoods. Biodiversity and Conservation 16: 2401-2427.

Bierregaard, Jr. R. O., Gascon, C., Lovejoy, T. E. & Mesquita, R. 2001. Lessons from Amazonia: The Ecology and Conservation of Fragmented Forest. Yale University Press, London. 478pp. BirdLife International (2000). Threatened birds of the world. Barcelona and Cambridge, UK: Lynx Edicions and BirdLife International. 852pp.

Birdlife international 2008, Bird conservation international (2008): Implications for deforestation for the abundance of restricted-range bird species in a Costra Rican cloud-forest,

BirdLife International. (2003). BirdLife's online World Bird Database: the site for bird conservation.Version2.0.Cambridge, UK: BirdLife International. Available: http://www.birdlife.org (accessed 4/12/2003)

BirdLife International. (2016). Pternistiscamerunensis. The IUCN Red List of ThreatenedSpecies2016:e.T22688340A93193443.http://dx.doi.org/10.2305/IUCN.UK.2 0163.RLTS.T22688340A93193443.en

BirdLifeInternationaleand nature Serve (2014). Bird species distribution maps of the world.2012. *Phylloscopussibilatrix*. The IUCN red list of threatened species. Version 2015.2

BirthLifeInternational. (2000). Threatened Birds of the World. Barcelona and Cambridge, UK: Lynx Edicions and BirdLife International.

Borrow, N and Demey, R. **(2004),** Field guide to the birds of western Africa. Christopher Helm, London. 510pp

Boudjemad, K., Lecomte, J. and Clobert, J. (1999). Influence of connectivity on demography and dispersal in two contrasting habitats: an experimental approach. Journal of Animal Ecology 68: 1207-1224.

Brooks, T. M., Pimm, S. T., Kapos, V. and Ravilious, C. 1999. Threat form deforestation to montane and lowland birds and mammals in insular Southeast Asia. Journal of Animal Ecology 68: 1061-1078.

Burgess, R. L. and Sharp, D. M. 1981. Forest island dynamics in man dominated landscapes. SpringerVerlag, New York, U.S.A. 239pp

Carole and Patrick Triplet (2012): Manuel de Gestion des Aires Protégées d'Afrique Francophone.

Central African Regional Programme for the Environment (2006). The forest of the Congo basin. State of the forest (http://carpe.umd.edu/documents, accessed 27/03/2013).

Cerutti, O., V. Ingram, and D. Sonwa. 2009. Cameroon's forests in 2008. In Les forêts du bassin du Congo: État des forêts 2008, eds. C. De Wasseige, D. Devers, P. De Marcken, R. Eba'a Atyi, R. Nasi, and P. Mayaux, 45-59. Luxembourg: Office des publications de l'Union Européenne.

Chace, J. F., and J. J. Walsh (2006). Urban effects on native avifauna: A review. Landscape and Urban Planning 74:46-69.

Cibois, A. 2003. Mitochondrial DNA phylogeny of babblers (Timaliidae), Auk 120: 35-54. Seth of Rabi Flickr(2012): Montane forest edge Becheve Reserve, Obudu Plateau, SE Nigeria: 26th june 2012.

Collar, N. J. and Stuart, S. N. 1985. Threatened Birds of Africa and related Island. The ICBP/IUCN Red Data Book. International council for Bird Preservation (ICBP) and international Union For Conservation of tne Nature and Natural resources,Cambridge,UK and Gland, Switzerland. 761pp.

Collar, N. J. and Robson, C.2007.family Timaliidae(Babblers) pp.70-291 in; del Hoyo., Elliott. And Christie, D.A. eds. Handbook of the birds.vol 12. Picathartes to Tits and Chickadees. Lynx Edicions, Barcelona.

Danjuma, D.F., Mwansat, G. S. Manu, S. S: The effects of forest land use and fragmentation on White-Throated Mountain Babblers(Kupeornis gilberti) a globally threatened bird species on the Obudu plateau, Southeast Nigeria: Ethiopian Journal fo Environmental Studies & Management 7 Suppl: 765-779; 2014

Dhindsa, M.S., Saini, H. K., Saini, M. S. and Toor, H. S. 1995. Food of jungle Babbler and common Babbler: A comparative study. Journal of the Bombay Natural History Society 92 (2): 182-189.

Dowsett-Lemaire, F. &Dowsett, R. J. (2000). Further biological surveys of birds in Cameroon.

Duveiller, G., P. Defourny, B. Desclée, and P. Mayaux. 2008. Deforestation in Central Africa: Estimates at regional, national and landscape levels by advanced processing of systematically-distributed landsat extracts. Remote Sensing of Environment 112: 1969-1981.

Elong, J.G. (2011). The urban elite in the Société de Développement du Cacao(SODECAO) cocoa revival project in forested Cameroon. In J.G. Elong, (ed.). L'élite urbaine dans l'espace agricole africain: exemples Camerounais et Sénégalais.Harmattan, Paris, 85-93

EU. 2009. Study on Understanding the Causes of biodiversity Loss and the Policy Assessment Framework. In: The context of the Framework contarct No. NG ENV/G.1/FRA/2006/0073. Specific Contract No. DG.ENV.G.1/FRA/2006/0073.

Eva, H., S. Carboni, F. Achard, N. Stach, L. Durieux, J.-F. Faure, and D. Mollicone. 2010. Monitoring forest areas from continental to territorial levels using a sample of medium spatial resolution satellite imagery. ISPRS Journal of Photogrammetry and Remote Sensing 65: 191-197.

FAO 2015. Global Forest Resources Assessment 2000, Main Report. FAO Forestry paper 140. FAO, Rome. Pp 562

Franklin, A. B. Noon, R. R. and George, T. L. 2002. What is the effect of fragmentation on the birds in western landscapes? Studies in Avian Biology 25: 20-29.

Franklin, S.E., Moskal, L.M., Dickson, E.E., Farr, D.R. & Hansen, M.J. 2000. Quantification of landscape change from satellite remote sensing. For. Chron. 76: 877-886.

Fuller, R. A.; Carroll, J. P.; McGowan, P. J. K. 2000. Partridges, quails, francolins, snowcocks, guineafowl, and turkeys. Status survey and conservation action plan **2000-2004.** IUCN and World Pheasant Association, Gland, Switzerland and Cambridge,UK.

GFRA (Global Forest Resource Assessment), 2010. The global forest Ressource Assessment

Gill, F and D Donsker(Eds).2013. IOC World Bird database. Lepage, D.2013

Gilliard, E. T. 1958. Living birds of the world. Hamish Hamilton. London.

Gottschalk, T.K., Huettmann, F. & Ehlers, M. (2005). Thirty years of analysing and modelling avian habitat relationships using satellite imagery data: a review. Int. J. Remote Sens.26: 2631-2656.

Gove, A. D., K. Hylander, S. Nemomisa, and A. Shimelis (2008). Ethiopian coffee cultivation-Implications for bird conservation and environmental certification. Conservation Letters 1: 208-216.

Grubbler A. (1990) Technology in the earth as transformed by human action eds B .L Turner, II, W.C Clark, R. W. Kates, J.E. Richards, J. T Mathews,

Hansen, M. C., Potapov, P. V., Moore, R., Hancher, M., Turubanova, S. A., Tyukavina, A., Townshend, J. R. G. (2013). High-resolution global maps of 21st-century forest cover change. Science, 342, 850-853.

IUCN (2014). IUCNRed list. *Downloaded from http://www.iucnredlist.org on 26/03/2014http://www.redlist.org/.*

IUCN, (2006). 2006 IUCN Red List of Threatened Species. Available from http://www.iucnredlist.org. On 02/06/2012.

IUCN (2016) The IUCN Red List of Threatened Species. Version 2016-3. Available at: www.iucnredlist.org.(Accessed: 07 December 2016).

Jansen, J.M.L., M. Bagnoli, and M. Focacci. 2008. Analysis of land-cover/use change dynamics in Manica province in Mozambique in a period of transition (1990-2004). Forest Ecology and Management 254: 308-326.

Johnston, C.A. 1998. Geographical Information Systems in Ecology. Oxford: Blackwell Science.

Kerr, J.T. & Ostrovsky, M. 2003. From space to species: ecological applications for remote sensing. Trends Ecol. Evol. 18: 299-305.

Kirby, K. R., Laurance, W. F., Albernaz, A. K., Schroth, G., Fearnside, P. M., Bergen, S., & Venticinque E. M. (2006). The future of deforestation in the Brazilian Amazon. Future 38: 432-453.

Laird, S.A., Awung, G.L. &Lysinge, R.J. (2007).Cocoa farms in the Mt Cameroon Region: biological

Lambin EF, Geist HJ, Lepers (2003). Dynamics of land-us and land-cover change in tropical regions. Ann. Rev.Environ. Res. 28: 205-241.

Laurance, W.F. (2006). Have we overstated the tropical biodiversity crisis? Trends Ecol. Evol. 22: 65-70.

Lepers, E., E.F. Lambin, A.C. Janetos, R. Defries, F. Achard, N. Ramankutty, and R.J. Scholes. 2005. A synthesis of information on rapid land-cover change for the period 1981-2000. Bioscience 55: 115-124.

Linder, J.M. Oates, J.F. (2011). Differential impact of bushmeat hunting on monkey species and implications for primate conservation in Korup National Park, Cameroon. Biological Conservation 144: 738-745.

Loveland, T. R., & Dwyer, J. L. (2012). Landsat: Building a strong future. Remote Sensing of Environment, 122, 22-29. http://dx.doi.org/10.1016/j.rse.2011.09.022

Malcolm L. Hunter, Jr AND James Gibbs: Fundamentals of Conservation biology; third Edition, 2007.

Marcel Holyoak: White throated-mountain Babbler (*Kupeornis gilbert*) Bakossi Highlands,Cameroon, 2012-04-05...102.jpg

Mariano Paracuellos: effects of long-term habitat fragmentation on a wetland bird community: 1 Aquatic Ecology and Aquiculture Research Group, University of Almería, Apdo. 110. E-04770, Adra, 4dlmería: Revue. Ecologie. (Terre Vie), vol. 63, 2008.

Marie Caroline Momo Solefack1, André Ledoux Njouonkou2, Lucie Félicité Temgoua3, Romuald Djouda Zangmene1, Junior Baudoin Wouokoue Taffo1 & Mama Ntoupka4(2018): Land-Use/Land-Cover Change and Anthropogenic Causes Around Koupa Matapit Gallery Forest, West-Cameroon: Journal of Geography and Geology; Vol. 10, No. 2; 2018

Mather, A. S., & Needle, C. L. (2000). The relationships of population and forest trends. Geographical Journal, 166, 2-13.

Mayaux, P., P. Holmgren, F. Achard, H. Eva, H.-J. Stibig, and A. Branthomme. 2005. Tropical forest cover change in the 1990s and options for future monitoring. Philosophical Transactions of the Royal Society of London B Biological Sciences 360: 373-384.

Meeta Kumari,Science Reporter, January 2013

Miavita (2011). Report on Mount Cameroon Socio-economic vulnerability and resilience. Collaborative project - FP7-ENV-2007-1.w

Millington, A., Walsh, S. & Osborne, P.E. (eds) 2002. GIS and Remote Sensing Applications in Biogeography and Ecology. Dordrecht: Kluwer Academic.

MINEF (1994). Law No. 94/01 of 20 January 1994 to lay down forestry, wildlife and fisheries regulations.

Mukete Beckline a, Sun Yujuna, Daniel Etongob, Sajjad Saeeda, and Abdul Mannanc(2018): Assessing the drivers of land use change in the Rumpi hills forest protected area, Cameroon: journal of sustainable forestry. https://doi.org/10.1080/10549811.2018.1449121

Munu Thomas Waithaka: An assessment of the impact of land use changes on human-elephant conflict in Laikipia West District, Kenya: A thesis submits in partial fulfillment for the degree of Master of Environnmental Science in the school of environnmental Studies of Kenya University, May, 2010

Murcia, C. 1995. Edge effects in gragmented forest: Implication for conservation. Tree 10: 58-62

Nagendra, H. 2001. Using remote sensing to assess biodiversity. Int. J. Remote Sens. 22: 2377-2400.

Otukei, J.R. (2006). Multi-temporal analysis of the multi-spectral landsat imagery for land cover change assessment (Case study of the Bwindi Impenetrable National Park), PGD, RS/GIS dissertation, ARCSSTEE, ObafemiAwolowo University, Ile-Ife, Nigeria.Secretariat of the Convention on Biological Diversity. Global Biodiversity Outlook 2. Montreal

Oyono, P.R., C. Kouna, And W. Mala. 2005. Benefits of forests in Cameroon. Global structure, issues involving access and decision-making hiccoughs. Forest Policy and Economics 7, no. 3: 357-368.

Patrick Triplet (2012): Manuel de Gestion Des Aires Protégées d'Afrique Francophone.

Pekin BK, Pijanowski BC (2012) global land use intensity and the endangerment status of the mammals species. Divers Distrib 18:909-918

Pickett, S. T. A. and Thompson, J. N. (1978). Patch dynamics and design of nature reserves. Biological Conservation 13: 27-37.

Pullin, S., A.(2002). Conservation biology. Cambridge University Press 359p.

Ralph, C.J. Sauer, J.R., and Droege, S. (1995). Monitoring Bird Populations by Point Counts. www.rsl.psw.fs.fed.us/projects/wild/gtr149/gtr_149.html.

Ranta, P., Blom, T.? Niemela, J.? Joensun, E. and Siitonen, M. 1998. The fragmented Atlantic rain foest of Brazil: size, shape, and distribution of forest fragments. Biodiversity and conservation 7: 385-403.

Republic of Cameroon, (1994). Law No. 94-01 of January 1994 To lay down Forestry, wildlife and Fisheies Regulations.

Rolstad, J. 1991. Consequences of forest fragmentation on the dynamics of bird populations: conceptual issues and evidence. Journal of the Linean Society of London, series B , 249-263

Rönkä, M., Tolvanen, H., Lehikoinen, E., Von Numers, M., Rautkari, M. (2008). Breeding habitat preferences of 15 bird species on south-western Finnish archipelago coast: Applicability of digital spatial data archives to habitat assessment. Journal of Biological Conservation 141: 402-416.

Rouse, J., Haas, R., Schell, J., Deering, D. & Harlan, J. 1974. Monitoring the vernal advancement of retrogradation of natural vegetation. Final report. Greenbelt, MD, USA, 371p.

Roy, P.S. & Tomar, S. 2000. Biodiversity characterization at landscape level using geospatial modelling technique. Biol. Conserv. 95: 95-109.

Sala Oe, Chapin Js, Chiozza F et al (2008) the status of the world's land and marine mammals: diversity, threat, and knowledge.Science 332:225-250

Sala, O.E., Chapin Iii, F.S., Armesto, J.J., Berlow, E., Bloomfield, J., Dirzo, R., Huber-Sanwald, E., Huenneke, L.F., Jackson, R.B., Kinzig, A., Leemans, R., Lodge, D.M., Mooney, H.A., Oesterheld, M., Leropoff, N., Sykes, M.T., Walker, B.H., Walker, M., & Wall, D.H. (2000). Global biodiversity scenarios for the year 2100. Science 287: 1770-1774.

Schmidt-Soltau, K. (2003). Rural livelihood and social infrastructure around Mount Cameroon. Mount Cameroon Project, Buea, Cameroon.

Scott, J.M., Davis, F.W., Csuti, B., Noss, R., Butterfield, B., Groves, C., Anderson, H., Caicco, S., D'erchia, F., Edwards, T.C. Jr, Ulliman, J. & Wright, R.G. 1993. Gap

Secretariat of the Concention on biological Diversity-SCBD (2006). Handbookof the convention on biological diversity including its Cartagena protocol on bio-safety, 3^e edition, Montreal, Canada

Sekercioglu, C. H. (2002). Effects of forestry practices on vegetation structure and bird community of Kibale National Park, Uganda. Biological Conservation 107:229-240.

Senapathi,D.,Vogiatzakis,I.,N.,Jeganathan,P.,Gill,J.,Green,R.,E.,Bowden,R.,G.,C.,Rahm ani,R.,A., Norris, K. (2007).Use ofremote sensing to measure change in the extent of habitat for the critically endangered Jerdon's Courser Rhinoptilusbitorquatusin India.Journal compilation © **2007** British Ornithologists' Union.149: 328-337

SHIIWUA MANU, INAOYOM SUNDAY IMONG And WILL CRESSWELL: Bird species richness and diversity at montane important Bird Area(IBA) sites in south-eastern Nigeria. Bird Conservation international(2010)20:231-239

Sibley, C. G. And Marone, B. L, Jr. 1990. Distribution and Taxonomy of Birds of the World. Yale University Press. New Haven

Stork, N.E., J.A. Coddington, R.K. Colwell, R.L. Chazdon, C. W. Dick, C.A. Peres, S. Sloan, And K. Willis. 2009. Vulnerability and resilience of tropical forest species to land-use change. Conservation Biology 23: 1438-1447.

Takem-Mbi, B. M. (2013). Assessing forest cover change in the Bafut-Ngemba Forest Reserve (BNFR), North West region of Cameroon using remote sensing and GIS. International Journal of Agricultural Policy and Research Vol.1 (7), pp. 180-187 September 2013, Available online at http://journalissues.org/ijapr/.

TakemMbi, B.M., and Roger, N. (2015). Cross River Gorilla (Gorilla gorilladiehli) conservation in a changing landscape of the Lebialem-Mone forest, South West Region, Cameroon. Syllabus Review 6 (1), 2015: 127 - 153

Tamungang, S. A., Cheke, R. A., Mofor, G. Z., Tamungang, R.N., and Oben, F.T. (2014). Conservation Concern for the Deteriorating Geographical Range of the Grey Parrot in Cameroon. International Journal of Ecology. Volume 2014 (2014), Article ID 753294, 15 pages. Available online at http://dx.doi.org/10.1155/2014/753294.

Titeu, N., Henle, K., Mihoub, J.B., Regos, A., Geijzendorffe, I., Cramer, W., Verburg , P., and Brotons, L. (2016). Biodiversity scenarios neglect future land-use changes. Global Change Biology, doi: 10.1111/gcb.13272

Toyi, M.S, Barima Y.S.S, Mama, A., André, M., Bastin, J-F., De Cannière C., Sinsin, B., Bogaert, J. 2013.Tree Plantation will not compensate Natural Woody Vegetation Cover Loss in the Atlantic Department of Southern Benin. Tropicultura, 31: 62-70.

Trumper, K, Bertzky, M, Dickson, B. , Van Der Heijden, G, Jenkins, M, Manning, P. (2009). The Natural Fix? The role of ecosystems in climate mitigation. A UNEP rapid response assessment. United Nations Environment Programme, UNEP6WCMC, Cambridge, UK.

IUCN/PAPACO. (2009). *Natural World Heritage in West Africa: status, conservation values*

Vicencio Oostra, Laurens G.L; Games and Vicent Nijman: Implications of deforestation for the abundance of restricted-range bird species in a Costa Rica.

Vie, J. C, Hilton-Taylor, C. And Stuart, Sn. Eds (2009). Wildlife in a changing world-an analysis of the 2008 IUCN Red List of threatened species. Gland, Switzrland: IUCN. Walsh, P.D., Tutin, C.E.G.? Oates, J.F.? Baillie, J.E.M, Maisels, F., Stokes, E.J., Gatti, S., Bergl, R.A., Sunderland-Groves, J. and Dunn,A. (2008). Gorilla gorilla. In: IUCN 2012. IUCN Red List of the threatened Species. Version 2012.1.(assessed on 22 August 2012, from www.iucnredlist.org).

Wafo, T.G., M.T. Demaze, And J.-M. Fotsing. 2006. L'information spatialisée comme support d'aide à la gestion des aires protégées au Cameroun: Application à la réserve forestière de Laf-Madjam. Paper presented at the Colloque international " Les interactions Nature-Société, analyse et modèles ", 3-6 Mai, in La Baule, France.

Walters, M. 1994. Bird's eggs. Dorling Kindersley. London

Wang, X.L., Bao, Y.H. 1999. Study on the methods of land use dynamic change research. Program Geography, 18: 83-89.

Wilcove, D. S.,Mclellan, C.H.& Dobdon, A. P. 1986. Habitat fragmentation in the temperate zone, pp237-256. In: Conservation Biology: the science of scarcity and diversity. Sinuaer, Sunderland, MA.

Wilcox, B. A. 1980. Insular ecology and conservation. - In: soulé, M. E and Wilcox, B.A.(eds), conservation biology: an evolutionary-ecological perspective. Sinuaer, Sunderland, MA? pp. 95-117

William J. Sutherland, Iamn Newton, Rhys E. Green: Bird Ecology And Conservation. Oxford Biology: A handbook of technique p42- 2005

Williams, M. (1990). "Forests" in the earth as transformed by Human action eds B.L Turner, II, W.C. Clark, R.W. Kates, J.E. Richards, J.T Mathews, and W.B. Meyer, 179-201 Cambridge: University press Cambridge, United Kingdom

Zogning, A; Claudia S. Ngouanet, C; Tchoudam D; Thierry P. Bignami C; Kouokam E; Buongiorno M.F; Ilaria M: Risks Prevention Plans(RPP) and the vulnerability of the peoples around volcanoes: Case study of Mt Cameroon 2010.

APPENDICES

Table: Mann-Whitney U Test (w/ continuity correction) Percceptions Marked tests are significant at p <,05000

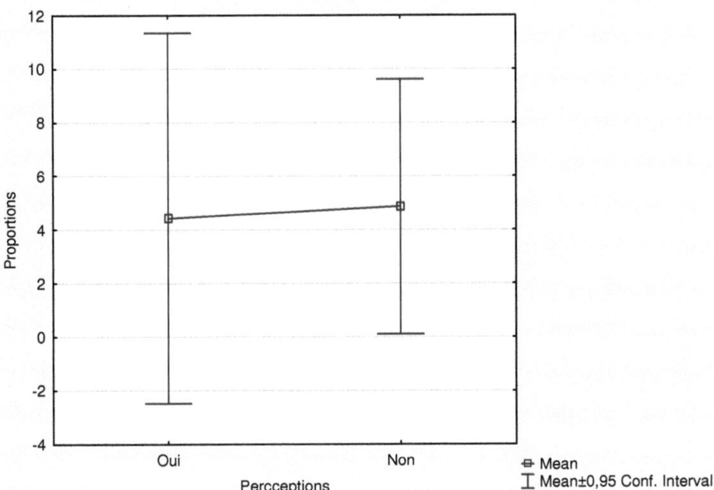

Diagram of proportions according to respondents' perceptions.

APPENDIX 2

APPENDIX 3: Field images

Example of forest conversion to cocoa plantation at Deck

Logging in Bakossi National Park

Kodmin secondary forest landscape

Savannah landscape in Kodmin

Species data collection walk

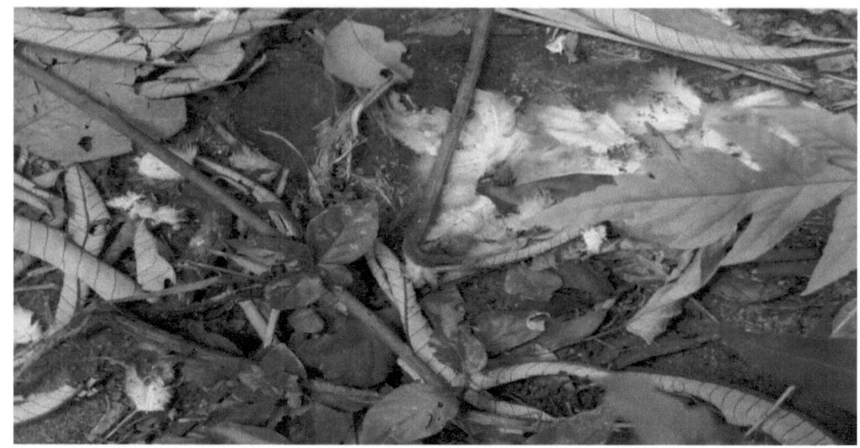

Bird feathers

QUESTIONNAIRE FOR THE LOCAL POPULATION

This questionnaire is framed purposely in order to get opinion from the local population concerning the species in our academic research work, the White throated-mountain Babbler (*Kupeornis gilberti*). The responses received from the interviews will be of private regard to each partaker. No person will be under obligatory to partake in this interview without the consent, in filling this questionnaire in order to avoid ambiguity, which will enable us to obtain good result, for the benefit of doubt, to better maintain its population for long-term sustainability.

Name: .... date..../..../201....localization ....hour:

Address: .... sex: ....nationality: ....

Region of origin: .... occupation: ....

1. Do you know the white-throated mountain Babbler (*Kupeornis gilberti*)?

a) Yes............... b) No.............

2. Have you ever seen this bird?

a) yes .... b) No ....if yes, where?

a) forest ....b) savana ....c) farmlands ....d) others... ......

3. How is call this bird in your dialect? ....

4. How is this bird important to you? ....

4. Do you eat this bird? a) yes ....b) No ....

If no why? ....

If yes how do you capture this bird? ....

5. How often do you see this bird? ....

6. a) Is there a particular season where the birds are more abundant?   a) yes...................

b) No................b) If yes which season?   a) rainy season ....

b) dry season........................

7. At what period of the day is the bird usually seen? a) morning......... b) midday...........

c) evening ....

8. a) according to you has the surface area of the forest reduces over time?

a) yes ……… b) no ………….. if yes farming: ………..Bush fire: ………population increase: …….
Timber exploitation: …… House exploitation: …… Fuel wood: ……

8. b) if yes which are the agents causing the decrease?

a) agriculture……… b) grazing……………..c) exportation by some organisations (which one)?…………………………..d) others…………………

9. How can you identify this bird in its habitat?

a) visual contact……………… b) vocal calls………………..c) others…………………..

10. On what does the bird feed? …………………

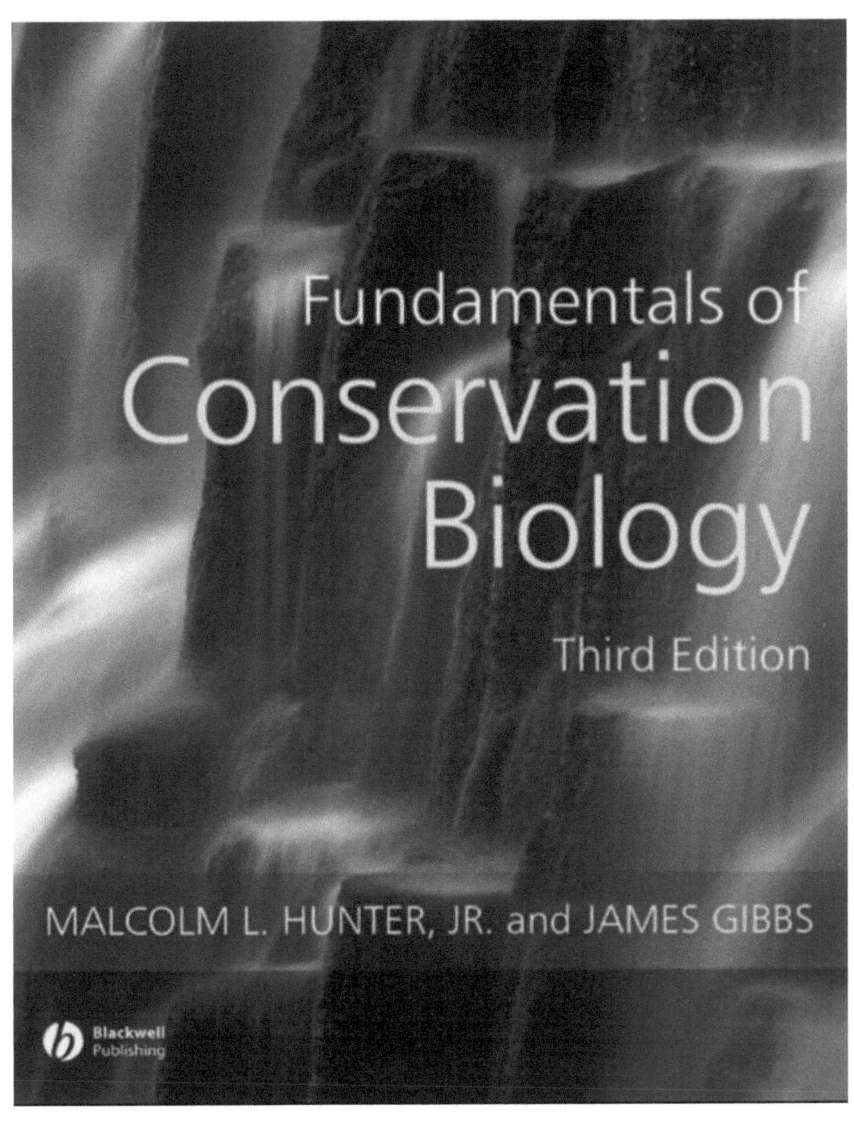

Fundamentals of
Conservation
Biology
Third Edition
MALCOLM L. HUNTER, JR. and JAMES GIBBS
Blackwell
Publishing

Printed by Books on Demand GmbH, Norderstedt / Germany